Nothing to Hide

Nothing to Hide

✦

Privacy in the 21st Century

Mark R. Keeler, Ph.D.

iUniverse, Inc.
New York Lincoln Shanghai

Nothing to Hide
Privacy in the 21st Century

Copyright © 2006 by Mark R. Keeler

iUniverse books may be ordered through booksellers or by contacting:

iUniverse
2021 Pine Lake Road, Suite 100
Lincoln, NE 68512
www.iuniverse.com
1-800-Authors (1-800-288-4677)

ISBN-13: 978-0-595-38843-1 (pbk)
ISBN-13: 978-0-595-83221-7 (ebk)
ISBN-10: 0-595-38843-4 (pbk)
ISBN-10: 0-595-83221-0 (ebk)

Printed in the United States of America

Contents

Acknowledgements

A book is rarely the work of one but rather of the many and there are those that must be thanked. In particular, special thanks go to Raquel and my family for their support and encouragement in everything I do. I would also like to extend my gratitude to a few special friends that gave me the inspiration and impetus to finish this book even if they don't realize how pivotal their support has been.

Fortis Fortuna Adjuvat

Preface

After spending several years researching privacy, even I must wonder what it is that drove me to write on a subject that can be incredibly dry, difficult to write, and of interest to only a few—and part of my reason for doing so was my own involvement in the development of technologies that could have irrevocably changed our perception of freedom and privacy. Years prior, while working on my doctorate, I had developed a next generation document protocol to be used with biometric passports that combined advanced cryptography, covert channels, and steganographic algorithms that was, by all accounts, a stunning achievement and I had graduated with distinction because of that work. But I was troubled by the potential that the potential presented a serious threat to privacy and civil liberties—and much of it was re-engineered to provide these protections.

That gave rise to many questions. What forces are causing the gradual erosion of privacy and civil liberties? Why is there such a gap between the masses that do not take their privacy very seriously and the few that take it too seriously? Is it a case of ignorance of the former or is it a case of paranoia of the latter? That is debatable but it remains that privacy offers a glimpse into the heart of society and provides a barometer of social health and the degree to which we respect others. And for that reason alone, it is something all of us should take to heart.

Of course there are varying degrees of privacy violations and the often—accidental questions asked by friends are generally benign compared to the more cynical attempts of organizations and governments to collect personal data on customers and citizens respectively. Privacy can be a very sensitive issue amongst family and friends, and despite best intentions, asking the wrong questions or providing the wrong answers in personal matters happens to everyone. But there is a profound difference between the best intentions of a friend and the self-serving interests of others and it is this expected benevolence, or lack thereof, that further divides those that can be trusted with personal information and those that cannot.

These kind of interpersonal interactions are far less malicious than the tactics employed by organizations or governments and only recently have they begun to address privacy issues—even when they are usually the worst of offenders. The majority of the population is somewhat cognizant of the fact that their privacy has been under assault for years, but the extent of that erosion, the extent to which everyone is subjected to some form of surveillance is only known in a few select circles. From the time you are born to the moment of ones deaths, from the time you awake to the moment you lay to rest, almost everything you do, say, and everything you've purchased is being scrutinized, examined, studied and observed.

Thus begins the war of privacy, a war that has been fought between two diametrically opposing factions under the surface for decades. Privacy is so much more than what you might think it is and there is much more at stake then you might realize. Far from just minding our own business, privacy is central to virtually every liberty we tend to take for granted, not the least of which is freedom. And yet despite this, there is clear evidence that the general public does not fully understand what privacy is and that is, perhaps, one of the central reasons that I chose this subject.

Unfortunately, of all the books available on the topic, and there are hundreds of books on privacy, the majority of those books tend to either focus on a single dimension of the problem, focus on a single audience, or simplify the problem to the point of where it loses its meaning and importance. Here too, particularly in the past few years, the numbers of books that focus exclusively on Internet Privacy, as if it were the only thing that mattered, have grown exponentially. This focus on the Internet, however, leads to a very myopic view of privacy that does not do justice to the subject.

"Nothing to Hide" takes a very different approach, one that to my knowledge has never been fully explored, and examines privacy, and by extension all civil liberties, as a function and product of *evolution* and written, as much as possible, for the casual reader. From the deepest philosophical paradoxes to social penetration theory and the internals of fingerprinting technologies, "Nothing to Hide" offers an alternative perspective on a timeless problem and provides readers with a wealth of knowledge that will change the way you think of privacy.

M. R. Keeler, Ph.D.

1

Foundations

1.1 Foundations of Privacy

"Democracy is the theory that the common people know what they want and deserve to get it good and hard."

—H. L. Mencken (1880–1956)

Let me start with the simple fact that privacy, though seemingly simple on the surface, is an inherently messy and complicated subject. Akin to the strong undercurrents of a calm lake, this is one of those subjects that is constantly shifting and adapting to a multitude of forces that are not readily apparent on the surface. Like the lake, privacy must not only keep pace with the lightening fast pace of technological change, it must also contend with constantly shifting social values, changing attitudes towards privacy and civil liberties, and a host of external forces such as government policy, and economics which in turn are influencing privacy or being influenced *by* privacy.

As a result, the subject requires a very broad multi-disciplinary approach that includes many of the hard and soft sciences; for example, practitioners have to understand the technical aspects of biometrics but they must be able to grasp the economic forces that are driving the technology forward, the sociological impact of surveillance as the technology becomes more prolific, and the effects on organizational structures that result from excessive surveillance.

Before you dive into the more complicated abstractions, perhaps it would not hurt to compare yourself with those whose privacy is constantly threatened–every celebrity and politician live the better part of their lives in a fishbowl and their experiences offer valuable lessons that will put privacy in a different light and improve your own practices as well. Suffice to say, if the details of their lives are fair game for a frenzied press, what makes anyone else's secrets any safer; indeed, from the Whitehouse to every "water cooler" in the world, there is nothing so sacred that cannot find its way onto a carnivorous grapevine that feeds the natural desire to learn about others. Most of us are neither wealthy nor famous and CNN is hardly pounding at the door but the same principles apply to everyone regardless of income, race, sex, or education. With few exceptions everyone, at some level, feed upon the flow of innuendo and gossip in both their personal and professional lives and the unfortunate proliferation of reality television, pandering to the lowest common cultural denominator, is but one indicator of deteriorating attitudes towards privacy and respect for others.

Taking the case of two co-workers that go to lunch more frequently that might considered the norm, even if there was nothing amiss, it only takes the wrong look or a misplaced word to unleash the torrent of office gossip. Who knows, maybe their kids go to the same hockey club, maybe they really are seeing each other—what business is it of ours? The same occurs when employees are leaving or transferring—once it hits the grapevine, there isn't a spin machine on the planet that can keep the truth from emerging. In situations like these, and it is a point that will be made later, gossip serves as a powerful and occasionally positive force. Mind you, gossip is not so amusing when *you* are the target of its relentless appetite and, in keeping with the subject at hand, how long would it take before *you* lament the loss of *your* privacy—what is good for the goose is not necessarily good for the gander, and when the situation is unmercifully reversed privacy suddenly becomes more important than ever before. If this is so, then why do we infringe on the privacy of others and demand respect for our own? Though I risk understating its complexity, it is nothing more than in the interest of self-preservation and an evolutionary need for communication. The human species have been sharing information since the beginning of time, and the acquisition of information is just one of many tools in our arsenal. And that bring me, loosely speaking, to the matter of biology.

Not surprisingly, every author is influenced by their education, experience, and personal convictions, and my academic interest in molecular biology has had a profound impact on the direction of this book; by extension, there is a tendency on my part to frame everything, especially open complex systems like privacy, in the context of evolution and I recognize this approach will not sit well with everyone. Every organism from the smallest, single cell species to the largest organizations in the world, seek to establish dominance over other species and you can find some pretty amusing parallels just by flipping between news and the Discovery channel—the only difference between the predator in the jungle and the one in the boardroom is their pinstripes though you always know where you stand with a Lion (lunch).

When it comes to predatory behavior, humans outperform any other species even if the façade of civility veils our more primitive desires; and part of this dominant pattern of behavior includes the acquisition of information that is required to make decisions regarding one's survival. Much of this explains why organizations go to such lengths to learn as much as possible about what individuals are thinking while simultaneously limiting disclosure about themselves. This is where privacy comes into play.

Privacy is nothing new and academics have been struggling with the concept for centuries—even Aristotle had once drawn the distinction between a person's private and public spheres as a classical reference to what information is or is not disclosed and this reference applies even more so today than it did during his time. If you were to view your life as a continuum from the most public to the most intimate, where do you draw the boundary of privacy? At what point does your life effectively become off-limits to discussion? This is far more problematic than it looks and the boundary tends to shift according to a volatile state of conditions and circumstances. Regardless of the complexities that define the concept of privacy, it is possible to simplify it as a set of unambiguous rights that clearly define our expectations.

The original sets of "rules" were drafted by William Prosser in 1969 and have become one of the most oft-cited works in the field. Let's begin by looking at the four definitive rights and then examine their relevance:

- *Intrusion upon a person's seclusion or solitude, or into his or her private affairs*

- *Public disclosure of embarrassing private facts about an individual*

- *Publicity placing one in a false light in the public eye.*

- *Appropriation of one's likeness for the advantage of another*

The first right in relation to a person's seclusion and private affairs is the foundation upon which the remaining rights exist. This right to seclusion in one's private matters has been expanded to include a broader range of protections beyond the usual the physical violations of space and trespass; it now covers emotions, thoughts, ideas, and various forms of human expression.

Decades old, the substratal premise of the first right provides protection against trespass of all private matters—what you are thinking, your diary, your personal life, and the decisions you make that concern your life—all of these things are covered explicitly under the first rule. There is sufficient evidence to suggest that the first rule has been eroded in the past 40 years to the extent that much of what was deemed private then has the potential to become public fodder; on this alone, there is reason to also suggest that society has not improved, it has regressed.

The second and third rules concentrate more specifically on the unauthorized and malicious nature of gossip and a relentless media that were best described in The Right to Privacy by Justices Warren and Brandeis in the Harvard Law Review of 1890; over two centuries later, the very same concerns they shared in 1890 have been magnified exponentially—they're concerns with respect to the printed press have superseded by 24 hour cable news and the Internet—but the underlying willingness to invade a person's privacy has remained a constant and it reinforces the argument that human nature hasn't changed all that much.

The last and Fourth of Prosser's Rules is particularly significant for the amount of the ground that it covers. In a more contemporary setting this may include the theft of one's identity, thoughts, aspirations, dreams, and ideas in addition to the appropriation of ones likeness in pictures as many celebrities have learned. And no less important, the Fourth Rule would include the trading of personal information to third party companies without explicit consent and which has been adopted within the framework of both Canadian and European privacy laws.

Though interesting from an academic and historical point of view, however, I cannot fault readers if they are inclined to wonder what relevance this has to everyday life. Bear with me a little bit and the importance of this will become crystal clear—privacy, as you shall soon discover, is about a lot more than who is reading your e-mail (that is a symptom but not a cause); on the contrary, privacy is about who controls your life. Everything you consider a personal and intimate concern, from doctor-patient confidentiality to discussions with your spouse and many of the things you routinely take for granted, are functions of privacy. And yet still, I am barely scratching the surface—the erosion of privacy affects not only what information is available to self-interested parties, but the decisions you can make without the coercive influence of those same parties. Consider a subject as contentious as abortion; putting aside the moral arguments for the moment, Roe v. Wade (1973) marked one of the most (in) famous applications of the right to privacy and it has become even more controversial since then. While social issues like reproductive choice have become celebrity causes of social conservatives, they epitomize the underlying importance of privacy—the battle for *control*.

Though somewhat reluctant to include a discussion of the Schiavo case for fear that it would appear insensitive but her plight exemplifies, at a very personal level, the significant of privacy and personal control and it can be discussed without disrespecting the family. To briefly recap, the Terri Schiavo case garnered national prominence in March of 2005 as the family and spouse fought to respect her wishes as she lay in permanent vegetative state. Regardless of your disposition towards the outcome of this incident, what was most alarming, not only to me but to the Supreme Court of the United States, was how quickly and shamelessly her plight was hijacked by Christian fundamentalists, anti-abortion activists, lobbyist groups, and politicians at every level; every one of these self-declared advocates jumped into the fray only in part for their concern for her well being but more to satisfy their own personal convictions.

And during the course of this incident, the vilification of the husband, numerous death threats, and the sheer willingness of the government and these parties to set aside the constitution in pursuit of their own agendas speaks to the issue of personal control over ones decisions—and the more socially contentious the decision, the more protection it demands. At stake then is not only the right to be left alone but the right to make fundamental decisions that have a bearing on the quality of life without interference or intrusion. At issue as well is the point at which decisions may be arbitrarily made for the "common good" and whose interests are being served. As a capstone to this unfortunate drama, U.S. republicans, at the time of writing, had begun to unleash veiled threats of retribution against the justices that presided over the case for not subscribing to the will of Congress—thus serving as a vital reminder of why the framers of the constitution sought to separate the three branches of government.

I was reminded of the old axiom that spirituality unites and religions divide, and it is firestorms like the aforementioned that heighten sensitivities and further polarize Americans with differing belief systems. Most disconcerting for me is not the following of any given religion but instead the insistence of its adherents that their beliefs must be superimposed on everyone else coupled with the superiority complex that provides a rationale for this behavior. The protection of privacy and civil liberties, including freedom of religion (which in turn includes the right not to practice a religion), becomes even more crucial when the choices of the few do not conform with the beliefs of the many. To make the point abundantly clear, enforcing the beliefs of the many over the few is called *oppression*—the very thing the United States has gone to war over on numerous occasions. For example, the first Amendment on freedom of expression allows me to claim that the sky is purple without fear of retaliation—it does not require that anyone believe the sky is purple, it does not require that you even read what I have to say—it only enshrines my right to make such statements—even when those statements are contrary to public sentiment let alone logic. These arguments do not constitute an attack upon any particular religion; instead it is to make the point that privacy provides protections to the individual when their belief systems do not coincide with others.

It can be quite tempting to throw the laptop and cellular phone off the balcony and hide under a rock but that isn't going to solve the problem; even if are not inclined to use technology with the same abandon as everyone else, the erosion of privacy still affects you. Nor will the problem be solved by halting the progress of science just because there are those that are uncomfortable with the direction it has taken. The privacy issue can only be solved by constructing a framework that guards personal information without unduly harming the economy or the progress an evolution of the species. Key to this framework is public education and hopefully this modest contribution will be your first step on that journey.

1.2 Philosophical Foundations

"I think we ought always to entertain our opinions with some measure of doubt. I shouldn't wish people dogmatically to believe any philosophy, not even mine."

—Bertrand Russell (1872–1970)

This can be a confusing thing as it is but adding to that chaos is the diversity of opinion and subjective interpretations that can be found amongst academics, lawyers, policy makers, and self-interested groups that often resembles the squabbles between relatives at Christmas dinner. On the one hand there is a school of thought known as Reductionism that is highly critical of privacy as a meaningful social or legal construct and on the other hand there is a diametrically opposite camp known as Coherentists that defends privacy as a meaningful principle. These are only schools of thought but it does illustrate the diversity of opinion with respect to privacy and getting acquainted with some of the different thinkers adds a new dimension to the subject.

As remarkable as it may sound, not everyone "buys into" the concept of privacy and it has been assaulted on the basis of numerous arguments, not the least of which is the belief that it is insufficiently grounded in legal doctrine, that it is economically inefficient, that it is not in itself a true concept but rather derived from more basic rights, and that it is actually *unnecessary*. The majority of scholars in the field take the opposing position that privacy is a fundamental necessity in order for civilization to mature and continue and I am clearly of this mindset. I haven't even begun to get into the special interests of government and the private sector that has their own agenda—thus far this is just debate amongst scholars. There is no doubt that there exists a clear distinction between the domains of your life that are either public or private; a person's thoughts, fears, emotions, dreams, aspirations—these are not the things destined for public consumption and review. Academics on either side of the debate must surely accept that there are some things best left unknown and it does raise the question of how they perceive their own privacy.

Returning to the philosophical roots of privacy, the easiest path to comprehending the divergent ideas is to look at the divisions between the prevailing theories of Reductionism and Coherentism; there are other theories that focus on the constitutionality of privacy and similar esoterica but these are the big guns on the subject. That is not to imply that Reductionists reject privacy, instead they argue that privacy is already covered under the basic liberties such as independence and autonomy.

The most noted of the Reductionists, Judith Jarvis Thompson, argued that privacy isn't actually a right or value in and of itself but instead is based on or derived from more elementary rights—in other words privacy can be reduced or broken down to rights and values that already exist. Thompson, referring to privacy as a "cluster" of rights (I prefer Aggregate Rights), argues that privacy is reducible to basic rights such as the right of private property, the right to be secure from personal harm, and other civil liberties and that privacy is nothing more than an umbrella that encompasses multiple social values that are already protected. More academics like Posner and Bork offer different views on the reducibility of privacy along constitutional and economic lines.

Posner's positions are comparatively, well, harsh to say the least and based *purely* on economics with what appears to be very little regard for the human dimensions of privacy; in his worldview, privacy is seen as a negative, with the assumption that information concealment is an act of deception and omission and, in an even more extreme position, argues that privacy does not enhance wealth (whose?) and lastly *that corporate privacy is more crucial than personal privacy* because people do not enhance the economy. I have to wonder how popular he would be at the annual Christmas party.

There is also the work of Bork that focuses more narrowly on the constitutional merits of privacy or more precisely the disdain he holds for the Supreme Court, specifically Chief Justice Douglas, for their recognition of privacy in 1965 in Griswold v. Connecticut. His principal position is that the Supreme Court should have not waded into what was ostensibly a social issue, that the court was actually creating a new law rather than interpreting existing law as they are chartered to do, and finally that their decision was not based on pre-existing legal doctrine within the framework of the constitution but the Supreme Court has had a long history of wading into social issues such as abortion and the right-to-die, so his argument, in my opinion, is without merit.

The debate as to whether privacy is or is not meaningful begins to wear on the nerves after awhile and there is no question that privacy *is* meaningful as a core value that a vast majority of North America's population consider integral to their peace and personal security. For all of the deep and eloquent passages, the witty constitutional arguments, all of that, the fact remains privacy is about *control*, that basic right to control the circumstances under which information can be released and disseminated. Given this apparent distinction between what is public and what is private, it is necessary to further define the difference between documented versus undocumented information.

Documented facts about your life are those that usually available to the public domain like your name, address, phone number, and simple identifiers—the information that we need to function properly in any society and that, for the most part, does not encroach upon intimate dimensions. For example, the courier that delivers your packages needs to know your name and address just as this information is often required for many transactions; and we often selectively share facts like where we go to school or where we work. There is what we refer to as the documented facts—things that may be referred to as public record.

Then there is the realm of the *undocumented*, including as it does the things that are not recorded and not available to the public—your thoughts, emotions, feelings, ideas, beliefs, political leanings, intimacies, all of these things are those that we want to protect the most. But as soon as *you* release such facts, as soon as they are not longer considered private, then they are documented facts and part of the public domain. That doesn't imply any of this would suddenly find its way to the company bulletin board, but there is the potential that it could. The distinction in this case is your willingness and desire to declare your preferences publicly for whatever reasons or motives you might have versus the unauthorized disclosure of private and intimate facts.

An interesting case in point is the U.S. Military's policies on sexual orientation that effectively state "Don't Ask, Don't Tell". Putting aside any discussion of discrimination or morality, the policy essentially means a soldier cannot admit to, acknowledge, or engage in conduct that would raise question of their sexuality for to do so is grounds, under military law, for a court martial and discharge. This raises two cases, one in which the soldier declares their sexuality and the other where the soldier behaves in a manner that provides evidence of their apparent preferences. In the first case, to publicly declare such intimate facts renders void any right to privacy because the facts were disclosed willingly and with full knowledge of the consequences. If, however, a solider did not make an overt statement but did frequent an establishment then where does privacy come into the equation—if the place is open to the public then the solider still made a conscious choice to enter the establishment and might be used as evidence at a court martial, even if the evidence is rather circumstantial.

Consider a situation that that hits a little closer to home and is far more applicable to the majority of readers in which a co-worker has decided to pursue employment elsewhere. Using the same criteria, until the employees declare their intent to depart or conducts themselves in a manner to suggest they are departing then this remains a private fact. If, however, the employees either declare they are leaving or were found to copying their resume at work then management is well within their rights to discharge the employee. Even if other employees know of their imminent jump they are still bound to honor the privacy of their co-worker.

None of this renders the theories any less significant nor does it minimize their contribution to the body of knowledge on privacy; even if I do not concur with Posner's philosophy, it does not invalidate his work. Likewise, the Schiavo case negates Bork's argument that the Supreme Court does not wade into social issues—it most certainly does. Indeed, and on a bright note to all of this, it ought to provide some degree of comfort that the justice system can work in favor of the individual even when the decisions of the court run counter to the wishes of the state.

1.3 Puzzles and Paradoxes

"...the paradox is the source of the thinker's passion, and the thinker with-out a paradox is like a lover without feeling, a paltry mediocrity"

—Kierkergaard

What exactly is a paradox and where do paradoxes come into any discussion on the subject of privacy? Let's start with the paradox. There are several definitions but the common theme is an assertion that is essentially self-contradictory though predicated on a valid deduction from acceptable premises; it can also be said that a paradox is any statement that asserts its falsity. Ergo, to say "I am lying" is a contradiction because in that statement I am telling the truth. Though this type of circular logic can be enough to have a person committed to an institution, I will share a few of them before getting into some of the social paradoxes related to privacy. There are all sorts of paradoxes like "I love you but I hate you", "It's sunny but it's raining", and then there are logical paradoxes like Russell's Paradox which poses the contradiction that if R is the set of all sets which don't contain themselves, does R contain itself? If it does then it doesn't and vice versa. And one of my favorites, an infinite is really a straight line.

Before I get into the paradoxes of sociology, here is one of the tougher ones to give you an idea of how vexing a paradox can be—you can skip this paragraph if your are prone to headaches. You start with the statement that an arrow in flight is really at rest. At each point in flight, the arrow occupies a length of space equal to its own length—it cannot populate a space that is either greater or lesser than its own length. But the arrow cannot move within this length that it occupies. It would need extra space in which to move, and it of course has none. So at every point in its flight, the arrow is at rest. And if it is at rest at every moment in its flight, then it follows that it is at rest during the entire flight.

Moving away from the abyss of obscure mathematical logic, there are also many social paradoxes and these are much more relevant to privacy—even if I take the scenic route to get you there. There are quite a few paradoxes that apply to individuals and society. For example, according to the Smarandache Social Paradox, to allow non-democratic ideas in a democracy infers the democracy will not survive. But to forbid such ideals means it is not a democracy. Even technology exhibits paradoxical behaviors. In the case of the Internet, you have a "social technology" that paradoxically diminishes social interaction by reducing the time spent in person. Equally contradictory is the fact that the Internet is simultaneously improving and harming participation in community life and social relationships.

On a closer inspection of many social phenomenon a series of seemingly intractable contradictions begins to emerge, and privacy, possessing of its own unusual contradictions, presents several paradoxes of human behavior. Contrary to what you might expect, privacy is by no means a static, monolithic abstraction. Instead it is a fluid, volatile reality constantly adapting and evolving in step with technology and social changes. Nor can privacy be fully examined without considering the psychological, sociological, and technological influences that in turn alter the meaning of privacy. This in turn leads to the discovery of some unusual behavior that makes privacy such a fascinating topic.

A utopian, simplistic worldview would insist that privacy was absolute and that private facts were yours and yours alone to maintain as you see fit. That same worldview would further restrict what could be asked of you in the first place. Alas, whilst the argument holds merit, observations of human behavior do not support the thesis. Indeed, such qualitative observations provide a landscape diametrically opposite that of any idealistic view of privacy.

It is time to introduce the first paradox.

The first paradox states that most of us claim to cherish our privacy and view it as a fundamental human right. And yet many of us will wantonly disregard our privacy in pursuit of a material gain.

From a quantitative perspective, every survey and study on consumer attitudes towards privacy indicate that an overwhelming majority of participants consider privacy to be one of the most important "rights" that they have. But it is not uncommon for that same overwhelming majority to sacrifice their privacy in exchange for things that are of questionable value such as the surrender of personal information in exchange for free gifts or discounts.

As for statistical data, there is no shortage of surveys but they all portray the belief that the population values its privacy. To quote just a few of them,

- *The March 2000 Business Week Privacy survey found that 86% of those polled want web sites to have their consent before collecting let alone sharing information.*

- *The August 2000 Pew Internet & American Life Project Poll showed that 86% of respondents supported opt-in privacy policies.*

- *The Time-CNN Poll indicated that 93% of respondents want companies to get their permission before selling personal information*

- *Less than 15% favor self-regulation of privacy according to the Business Week/ Harris Poll*

And this continues ad infinitum. The numbers tell a striking story of consumers that are angry and confused over their privacy rights and seek protections that will guard against abuses. This is by no means surprising and the results likely confirm your own sentiments and concerns towards privacy. Betraying the numbers, however, is a lot of evidence that, though more qualitative, illustrate the tendency to quickly surrender privacy when presented with material gains or where when prompted for personal information without challenging the legitimacy of the request. I've already encountered several occasions where attempts to pay by cash were thwarted by information systems that required personal information to complete the sale; and so entrenched was this culture that it was difficult if not possible to print an invoice. This is what you might refer to as a form of institutionalized or systemic intrusion of privacy.

The fact is, now more than ever, information is viewed by organizations to be more valuable than gold, and the more personal it is, the more it tells them something about your *preferences* and *behavior,* understanding your behavior makes it that much easier to sell to your preferences; but to get into a person's wallet requires that companies first get into their head to understand what drives their purchasing decisions. There are (quasi) academic fields, like consumer anthropology, that are dedicated to understanding consumer behavior—more specifically what forces influence consumer behavior. In many of the big box stores, psychologists and anthropologists spend considerable time observing movements through the store. They know how to get you in the store, keep you there, and while there, direct you through the store as they wish—they know where you will go, what direction you will turn, where you will look, and what you will pick up. After all, why do you think the first 15-30 feet of a big box store is so empty—that is a decompression zone, an area where shoppers adjust their eyes and orient themselves to their surroundings. Everyone, myself included, would like to think they are savvy to these tactics but the body, the face, our movements, all of these things convey information to the external observer that we not aware of. Once armed with this kind of data, when coupled with basic principles of social psychology, marketers have an uncanny ability of bringing consumers to the precipice of self-disclosure.

A good salesperson doesn't ask *if* you want the item, they ask *when* you want it delivered thus short-circuiting the decision tree that could result in a negative response. Likewise, good marketers know that you will be defensive until they can figure out what it is that motivates you—a chance for a free trip, discounts, appeals to your social integrity (self-perception), or anything that appeals to your ego. And there is nothing better than the reciprocal obligation of a "free gift". How do you think they know this? Through a combination of observation and statistical modeling of large segments of the population.

But the focus has now changed to targeted marketing (this is not new) in which the market attempts to market to individuals rather than larger demographic blocks. However, to do so obviously requires a more intensive (invasive) examination of consumers. Why do you think even the most gratuitous use of sex works so well—as one textbook stated sarcastically, every successful commercial uses cars, women, children, or dogs, or a combination thereof. Sex sells for the simple reason we are a sexual species. That may clash with the misconception that society has become more enlightened, but there are enough people that fall for it to justify its use repeatedly—there is a good reason why football teams have cheerleaders. That does not undermine the integrity of the women; it speaks to basic human nature.

For the sake of brevity, it suffices that the marketplace has developed very sophisticated techniques to determine personal preferences. And with this information, it becomes easier for organizations to pry information from consumers by applying the right pressure at the right time because they realize that the temptation of saving money or getting something for free is often enough to get consumers to give up a little privacy.

And that brings me to the second paradox…

The second paradox states that there is a dichotomy between the natural desire to tell all and the countervailing desire to maintain our privacy.

I don't want to steal my own thunder from Chapter Two but it bears mentioning that humans are, first and foremost, social animals and the sharing of information is both healthy and necessary. The problem, if I may call it that, is *what* information is being shared, *when* information is disclosed and with *whom* we share it. After all, the many times we've been frustrated by our inability or unwillingness to say what is really on our minds may be loosely described as psychological paradox between the need to constrain oneself (privacy) and the need to tell all (disclosure). Freud once wrote that there is a gap between what a man is (reality) and what a man wants to be (dreams)—for example, lots of people would like to be the President of the United States but fate shines kindly only on the privileged few. I've played around with this principle to extend it to privacy and disclosure and put forth that the "gap" between what we can say and want to say is proportional to the level of stress endured. I have my own favorites but will leave you to think of a few times when this applied to you. Somewhere along the continuum of disclosure and privacy, sits the barrier of privacy and each defines for themselves where that barrier begins even though it is constantly moving along this line in response to varying social circumstances, the target audience, and the topic of discussion; what I might consider very private at this very minute could be an open book in another context at another time and with another person.

Before I set off a firestorm in the next section by taking a swipe at social conservatives, the paradoxes that define human behavior are not that hard to comprehend; there is a conflict that exists between idealized behavior and the pragmatic choices that everyone have to make from time to time. It is difficult for human beings to remain chaste in their words because we are genetically predisposed to sharing information. Still, I like to come back to this belief that all systems require a point of equilibrium and it is my conviction that this can be applied to privacy and disclosure; there must be a way to achieve a healthy balance between open disclosure amongst trusted parties and a more reserved, cautious attitude towards non-trusted entities.

1.4 No Trespassing

"The protections of the Fourth Amendment are clear. The right to protection from unlawful searches is an indivisible American value. Two hundred years of court decisions have stood in defense of this fundamental right. The state's interest in crime-fighting should never vitiate the citizens' Bill of Rights."

—John Ashcroft, Chairman of the Senate Commerce Committee on Consumer Affairs, Foreign Commerce and Tourism, 1997

Unless you went to Harvard Law School the thought of studying constitutional law on a Friday night is probably not on your short list of things to try but then again you do not need to study property law to know that a person's home is his or her castle and our culture recognizes and reinforces this basic right. Even someone trampling through your garden instinctively knows they are not allowed in your backyard without your permission—and the law too recognizes the sanctity of the home and provides legal redress for abuse of your property rights.

The "right to privacy" has meant different things to different people and different things in different times. Scholar W. A. Parent considers the following to be the most common views of what the right entails: 1) the right to be left alone, 2) the right to exercise autonomy or control over significant personal matters, and 3) the right to limit access to the self. If you look closer you will notice that privacy entails a lot more than stepping on your lawn and really starts getting into the issue of personal freedom and the ability to make decisions that affect your own self-interests; when a person or group interferes with that process they your space a well.

To put things into perspective is best exemplified by examining two of the more contentious issues that have polarized the United States for decades—specifically abortion and pornography. For myself, personally I could care less what you watch or what you do behind closed doors or even open doors for that matter—but not everyone shares in my celebrated indifference to your affairs and to their way of thinking it *is* their business. It isn't as if this is a new or unique dilemma and history is replete with examples of one group that claims superiority over another that has often resulted in warfare. Differences in wealth, class, race, color, sex, and religion have long been the flashpoints of human conflict and much of that conflict stems from a failure to respect others and it is more often than not an exercise in futility trying to explain an opposing point of view to either faction. Both will prove stubborn inflexible, and set in their ways.

It is one thing to hold strong convictions and yet another to attempt to impose those beliefs on others, but the polarization of social issues in the United States point to a very different reality that pits the Christian Extremists against Liberals (for lack of better definitions) and there is every indication that the country is becoming more polarized rather than less. The term Religious Right refers to those groups that present a threat given their demands for dramatic social changes that are based on their own religious doctrines. What makes these positions so dangerous is not just their beliefs on specific social issues but their belief in who should hold power and, more importantly, how that power should be exercised.

> *"While it is true that the United States of America was founded on the sacred principle of religious freedom for all, that liberty was never intended to exalt other religions to the level that Christianity holds in our country's heritage."*

> **—Family Research Council**

> *"Congress shall make no law respecting an establishment of religion, or prohibiting the free exercise thereof; or abridging the freedom of speech, or of the press; or the right of the people peaceably to assemble, and to petition the Government for a redress of grievances."*

> **—First Amendment, United States Constitution**

Well unless my eyes deceive me, and they don't, the First Amendment of the United States Constitution in no way differentiates or gives preference to one religion over another and the statements made by the Family Research Council are yet another example of extremist Reconstructionism. Christian Reconstruction is a call to the Church to awaken to its biblical responsibility to revival and the reformation of society. While holding to the priority of individual salvation, Christian Reconstruction also holds that cultural renewal is to be the necessary and expected outworking of the gospel as it progressively finds success in the lives and hearts of men. Christian Reconstruction therefore looks for and works for the rebuilding of the institutions of society according to a biblical blueprint.

At issue is also the means by which they seek to impose those beliefs on everyone else. And one of the first priorities in this strategy would require the abolition of privacy and civil liberties. Well I could go on for hours but the point is this: any time you have one group that believes itself superior to any other, there is going to be an effort to control and dominate the other and history supports this. As a final measure of such extremism the words of Ann Coulter come to mind.

> *"We should invade their countries, kill their leaders and convert them to Christianity. We weren't punctilious about locating and punishing only Hitler and his top officers. We carpet-bombed German cities; we killed civilians. That's war. And this is war"*

—Ann Coulter

There is nothing to be served by paraphrasing statements like this when they have been made on the record and require no clarification from me—this is indeed a war of beliefs between those that want freedom and choice and those that want to make those choices for everyone else. That is not an assumption, it is a well documented and published account. Returning to the issue of pornography, to indulge or not is entirely a matter of choice and that speaks to the right to make such decisions without interference or fear of reprisals. It is possible that positions defending choice and privacy will be assaulted on the grounds that a defense of privacy is tantamount to a defense of pornography but this fallacious and inflammatory logic at best. As for abortion, the Supreme Court has made it abundantly clear interference in a woman's choice is a violation of her privacy and by extension a violation of the Ninth Amendment. Though privacy is not specifically protected, the court has seen fit to provide such protections—something that has long been a source of anger for opponents of abortion.

One of the greatest threats is the Pandora's Box that will be unleashed if privacy is to be set aside for some arbitrarily defined common good; assuming that pornography could ever be eliminated what stops the juggernaut from attempting to stamp out anything it deems offensive—perhaps we ought to eliminate the writings of Oscar Wilde and while we are at it, eliminate the First Amendment. The same is true of the release of medical information of *any* kind—once it has begun it can never be stopped and we may find ourselves facing a new era of eugenics in which those less fortunate are deemed unworthy by virtue of illness or infirmity. It is not much of a jump from publishing the names of women that have abortions (which is possible in certain states) and the publishing the name of anyone with an illness.

Given the variety of beliefs regarding the content of privacy rights, and the absence of an explicit reference to privacy in the Federal and in most state constitutions, it is no surprise that courts interested in protecting privacy have protected an array of interests in its name. For instance, the U.S. Supreme Court has within the scope of privacy protected child rearing, education, contraception, and reproductive rights. It has also considered the issues of peddlers going onto private property and disturbing homeowners, and loud trucks running through residential neighborhoods to involve privacy interests. The concept of privacy and limited privacy rights was recognized in ancient Athens. Indeed, the language, law, and writings of the period reveal that privacy and property in Athenian society were interconnected, and recognized as such. While Athenian law respected a form of privacy, Plato believed that privacy could not serve a constructive social or psychological purpose, and argued for the eradication of the private realm. He believed that wiping out private property would contribute significantly to the elimination of all that is private, including thoughts, emotions, desires, judgments, and decisions.

Plato's vision was never realized in Athens, nor was it influential in English and American jurisprudence, which continued in the Athenian vein to recognize and develop privacy rights in connection with property rights. In 18th-century England, the early parameters of what was to become the right to privacy were set in cases dealing with unconventional property claims. In Pope v. Curl (1741), a bookseller named Curl obtained and published, without consent of the authors, personal letters written to and by well known literary figures, including Alexander Pope and Jonathan Swift. Pope sued Curl, seeking to have the book containing the letters removed from the market and Curl enjoined from similar actions in the future. The Lord Chancellor upheld the privacy of Pope's letters on the grounds that the writer of a letter has a property right in his words.

In the 1820 case of Yovatt v. Winyard, the court extended property rights protections to cover personal secrets. In that case, Winyard, a journeyman assistant, left the employ of Yovatt, a veterinarian, to start a competing business. Winyard used secret medicines in his new practice, providing clients with printed instructions on how to use them. Yovatt sued, alleging that Winyard had obtained the formulas for the medicines as well as the instructions for their use from him by surreptitious and clandestine means. Particularly, Yovatt believed Winyard had copied the information out of his personal book. The Lord Chancellor ruled in Yovatt's favor on the grounds that there had been a breach of trust and confidence, and ordered Winyard to stop using the formulas and instructions.

Yovatt brings to light the interesting and important fact that what we now call 'unfair competition' and 'plagiarism' and 'privacy' were all wrapped together, in Yovatt's time, under the principle of 'property.' It was only later that these concepts were separated.

A third case that contributed to the development of privacy rights was Prince Albert v. Strange and Others, decided in 1849. The case was famous because the plaintiff was the husband of Queen Victoria, and the Queen herself was an aggrieved party in the suit. In dispute was the right of printer William Strange to sell reproductions of etchings that he had catalogued and printed without the consent of their creators, Queen Victoria and Prince Albert. While the right to privacy was not explicitly recognized at the time, Victoria and Albert argued their case in terms of their right to keep private art they had created for their personal enjoyment. Realizing that the court would protect a property interest, but not an independent privacy interest, Strange's lawyer sought to capitalize on the distinction. He observed, "It has been argued that privacy is the essence of property, and that the deprivation of privacy would make it, in fact, cease to be property." He concluded that "the notion of privacy is altogether distinct from that of property." The court did not accept his argument. Ruling in favor of Victoria and Albert, the Vice Chancellor wrote: "Every man has a right to keep his own sentiments, if he pleases. He has certainly a right to judge whether he will make them public or commit them only to the sight of his friends. In that state the manuscript is, in every sense, his peculiar property; and no man can take it from him, or make any use of it which he has not authorized, without being guilty of a violation of his property". According to one commentator, the most significant aspect of this case and its underlying philosophy is that it rested on a right of privacy, which the court considered a type of property right. In fact, it appears that until 1890, no English court recognized the right to privacy independent of property rights.

Across the Atlantic, the right to privacy was developing in a similar fashion, as an outgrowth of property rights. The Third, Fourth, Fifth, and Fourteenth Amendments to the United States Constitution all protected people and their property against government intrusions. In protecting property, these Amendments also protected privacy. "The legal maxim and popular proverb that 'a man's house is his castle' had wide application in the nineteenth century." Civil and criminal penalties threatened anyone who dared invade the sanctity of the home, or disturb the quiet possession of the householder. The Fourth Amendment's prohibition of unreasonable search and seizure, as well as the law of trespass, were viewed by courts as safeguards of a homeowner's privacy. Damages in trespass even included compensation for "invasion of privacy." The first United States Supreme Court decision interpreting the Fourth Amendment recognized an "indefeasible right of personal security, personal liberty and private property" against "all invasions on the part of the government and its employees of the sanctity of a man's home and the privacies of life."

A critical event occurred in 1890, dramatically altering the course of the development of the right to privacy, and giving birth to the current philosophical dichotomy between privacy and property rights. That year, the Harvard Law Review published an article by Samuel Warren and Louis Brandeis entitled "The Right to Privacy." In that article, the authors argued that many decisions granting relief on the grounds of invasion of property, such as defamation, breach of confidence, or breach of implied contract, were really based on a broader principle-the right to privacy. Although they acknowledged that privacy was already protected within the ambit of property rights, they argued the right to privacy ought to be recognized and protected separately.

Warren and Brandeis believed that as society became more civilized and technology advanced, rights also should evolve to protect new threats to human dignity and emotions, and to preserve propriety and decency. The article began the process of divorcing privacy from its historical and intellectual partner, property rights.

The result of this divorce has been a confused understanding of the origin of the right of privacy, as illustrated by the seminal Supreme Court decision in the area. In Griswold v. State of Connecticut, the Supreme Court construed the right to privacy as a "penumbra" formed by "emanations" from the First, Fourth, Fifth, and Ninth Amendments, but did not link privacy explicitly to property rights. Similarly, in decisions ranging from abortion to criminal search and seizure cases, the Court has loosened privacy protections from their property rights moorings. In the process, it has muddled the parameters of the right and allowed critics to argue that the right to privacy does not exist in the Constitution.

Ironically, the false dichotomy between property and privacy rights-and the need to join the two concepts together again-is perhaps best illustrated by the case of Moore v. City of East Cleveland. Moore involved the criminal prosecution of an elderly black woman who, by having two of her grandchildren (who were cousins) living with her, violated a local zoning ordinance limiting occupancy of residential dwellings to members of a single "nuclear" family.

A divided Court struck down the ordinance as unconstitutional. The plurality considered the ordinance a violation of the right to privacy, as protected by the Due Process Clause of the Fourteenth Amendment, noting that the Court has "long recognized that freedom of personal choice in matters of marriage and family life" is constitutionally protected.

Where Moore becomes interesting, however, is in the concurring opinion of Justice John Paul Stevens, who provided the swing vote in the five to four decision. Justice Stevens viewed the "critical question", as "whether East Cleveland's housing ordinance is a permissible restriction on [Mrs. Moore's] right to use her own property as she sees fit". Stevens observed, "Long before the original States adopted the Constitution, the common law protected an owner's right to decide how best to use his own property". In Stevens's view, the application of the ordinance constituted a "taking" without due process or just compensation, in violation of the Fifth Amendment. The Moore case illustrates the interconnectedness between privacy and property rights. Given the same set of facts, four members of the Court believed privacy rights were jeopardized, while another believed property rights were threatened. Ultimately, the two segments came together to protect the rights at stake.

Since 1977 and the Moore decision, the composition of the Court has changed, and a majority now exists that does not view favorably rights that are not explicitly defined in the Constitution. Those like Senator Biden who does not understand the nexus between property and privacy rights may unwittingly be creating the groundwork for the Court to diminish protection for privacy. The most enduring protection for both rights is to view each as indispensable to the other.

When asked about the Moore case, Judge Thomas replied that he agreed with the decision, and noted that his own family living arrangements as a boy in rural Georgia would have been unlawful under the East Cleveland ordinance. Thomas's personal experiences, growing up in an era of state-enforced segregation, likely gave him a keen appreciation for both property and privacy rights and for the consequences of denying such rights.

Out of the strange ritual that brought together the ideas of Clarence Thomas, Joseph Biden, Stephen Macedo, and Richard Epstein, among others, an important, although perhaps overshadowed issue emerged—what will be the destiny of privacy and property rights in the decades ahead? Interestingly, the answer may depend on whether the connection between privacy and property rights is rediscovered and acknowledged to be essential to our precious liberties.

To trespass onto the land of another, even when that property is not explicitly marked by signage denoting private property, still constitutes an offense in many jurisdictions where the law has ultimately determined that other markers or barriers are sufficient indicators to the offender that they have crossed a boundary onto the private property of another without the express permission of the owner. It is no less than if a person was to walk through the front door of your home uninvited. The same argument is easily manifest with respect to your personal privacy.

There are, when speaking of privacy, very clear boundaries over which no person has the right to intrude without the consent of the offended party. With that, there exists boundary that defines what personal information is available to the public and what information is considered private and personal. In many states and provinces, a person's name, address, and perhaps a phone number, which may be found in any telephone book, are the only attributes of the person that are deemed public knowledge; everything else is private. Thus to coerce the disclosure of such material facts or to seek the same data through alternate "infomediaries" constitutes a form of trespass.

The act of "spamming" now constitutes not only a criminal offence, there is reason to suspect that such acts also constitute an act of trespass, let alone breach of contract and breach of trust and the same can be said of Internet cookies, web dots, and many other technologies discussed later. It remains to be seen how well the law will respond to these violations and there is reason to expect, based on current case law, that these acts of invasion will be dealt with in both civil and criminal courts of law.

If a man walks across your lawn, he has clearly trespassed upon your private party and there are civil remedies to address this act of invasion; but is it any different if the same man asks questions of you or seeks information of a private nature that could cause more harm than mere crumpled grass? The grass will grow again and does not possess the ability to be offended—humans do. This book offers the reader the corollary that to trespass upon a person's private life is an invasion of our dignity and for that there are appropriate penalties to duly penalize and deter the offender thereafter. There are boundaries that should never be crossed.

Over the course of time, protection of our property has changed to the point where gated communities, barbed wire, and armed security are quite common; in other words, we've learned how to effectively defend the physical realm of our universe and that universe is well defined. But the same cannot be said of our defense of information rights, and privacy is frequently sacrificed without understanding what it is that has been lost until it is too late. Viewing personal and information privacy with the same territorial instinct reserved for the home is a first step in preventing invasions to begin with.

1.5 Nothing to Hide

"The power of hiding ourselves from one another is mercifully given, for men are wild beasts, and would devour one another but for this protection"

**—Henry Ward Beecher (1813–1887),
"Proverbs from Plymouth Pulpit", 1887**

That brings me to the final question of this chapter. Why did I choose to name this book "Nothing to Hide"? The expression "I have nothing to hide" and the opposite query "What do you have to hide?" is not a question or even a state of mind so much as it is a cultural shift in society that does not place an adequate value on privacy and related liberties. And yet, contrary to the axiom that we have nothing to hide, there are countless circumstances in which disclosure of private facts or thoughts can cause irreparable harm.

There are ample situations where sensitive facts, though by no means criminal, that can cause untold damage or embarrassment if revealed; and the argument that a person ought to reveal these facts is a coercive play on an individuals natural sense of indignation thus, the mere act of questioning a person's integrity is enough to provoke revelations that might otherwise be left unspoken. To put it another way, there are many elements of human behavior best served by secrecy including things like religion, sex, politics—the "Nothing to Hide" doctrine is nothing more than a form of social intimidation to expose those that do not conform. And equally disturbing than those that seek to extract information is the willingness with which information is so wantonly disclosed. This willingness to surrender information in exchange for returns of questionable worth or to conform to social norms is far more troubling because it signals an unwillingness to question the questioner or engage in critical thinking.

Undeniably, the degree of religious freedom found in North America is virtually unrivalled as noted by the more than 360,000 different institutions according to the ACLU; and that is, as it ought to be considering the fundamental civil protections provided by the Constitution. But there are still fractious divisions along polarized religious lines and governmental interference in religion can result in invasions of privacy as well as the curtailment of basic liberties.

There are irrefutable indications that, despite the perceived freedom of religion, there are varying degrees of religious intolerance and discrimination that can be found in many social circles, communities, the workplace, and government throughout North America. The number of Supreme Court cases that have been launched with respect to the posting of the Ten Commandments in public schools is just one prime example where one specific religion has acquired excessive influence and power to the detriment of all others.

When you add to this the proliferation of government funding of religious schools and the sudden explosion of faith based initiatives, it should be sounding alarms in your mind. Mind you there is merit to the argument that the United States was founded on Christian roots and that the posting of the Ten Commandments (or any religious symbol) is integral to American culture. The concern arises only when such practices come to dominate or even remotely influence a public institution—and given the sensitivity towards transparency and conflict of interest, this should really be a no-brainer—the government should always maintain an arms-length relationship from *all* religions.

Though there is no fault in embracing a particular religion, these seemingly innocent acts of faith lend themselves to an escalation of more overt discrimination, and there are cases before the courts as illustrated by the City of Omaha v. Hussein where a Muslim woman was denied entry to a public pool ostensibly because she was Muslim. To show a semblance of balance, there is also a case in Richmond, Virginia, in which the Cornerstone Baptist Church was prevented from performing baptisms in a public park by the Falmouth Waterside Park management.

Invasions of privacy can occur if organizations coerce an individual to reveal their religious faith; once that information is disclosed it is impossible to ensure that a person's choice of faith is not used to discriminate against them for employment, housing, healthcare, or any other service to which they are entitled.

And what happens when the questions begin to delve even deeper? What happens when employers move from questions to fluid samples? When it comes to drug testing and the disclosure of substance abuse problems, no matter the nature of the affliction, the very act of revealing that you have a substance abuse problem could well lead to termination or alienation from your peers—even if the abuse was many years ago.

To acknowledge a problem in the past may well lead to the assumption that it could happen again and, as a result, these disclosures can lead to financial ruin and damage to a person's social standing. I wouldn't want to prevent anyone from seeking the help that they need and if anything they ought to be commended for taking those first tentative steps—but I am somewhat skeptical that revelations of this nature will not cause considerable harm. In many industries, disclosure of said facts could severely limit a person's ability to find employment within their specialization.

And while it may come across more as a social commentary, the obsessive-compulsive focus on the "War on Drugs" leads me to believe that revelations of this nature are irreversible. There are already accounts of some companies banning and even terminating smokers and while their motivations may be based on financial considerations, there is still a subjective component that is passing judgments on these people for something that is still very legal. Even if you are a non-smoker, it is imperative to step away from the wrongs of smoking and focus on the act of discrimination based on disclosures. To extend the thought, being terminated for smoking could just as easily evolve into a termination for obesity.

Thus, to conceal such things is done to safeguard personal safety and security—not just physical security but also financial security and protection of one's reputation within a community or industry. This entire subject goes to the growth, and problems related, to drug testing in many American firms and it reaffirms the dangers of this "Nothing To Hide" mentality—and it presents a terrible Hobson's Choice, for to agree to drug testing is to surrender your privacy but to disagree infers that there is something to hide. No one should ever be compelled to make that choice.

I am reminded of a form of coercive logic that is often discussed as the following question. Have you stopped beating you wife? If you answer "No" then you must be beating your wife, but if you answer "Yes" then it still infers you have previously beat you wife (or whomever)—this form of question forces the individual into only two pre-determined answers even though they may have never committed this offence. The "Nothing to Hide" doctrine follows a similar path—if you do submit to a question then you have surrendered your privacy—and if you do not submit then you must be hiding something—either way there is no allowance given to challenging the question itself.

Since the "interrogator" (defined as anyone asking the question) is attempting to coerce harmful revelations it is only fitting that the motivations and the impartiality of the questioner in turn be called into question. Those in power are no less infallible and can easily be swayed by personal biases and their own self-interests—everyone is biased, and to argue otherwise runs counter to human experience. So in asking whether a person has anything to hide, it is incumbent upon the respondent to question the motives before answering the question. *Why* are they asking these questions? How will it be used against them? Who has access to the answers?

There is the risk that questions of this nature can grow, if left unchecked, to the point where the "Nothing to Hide" doctrine applies as much to the innocent as it does to the guilty. It is not a great leap of logic to ask where a person was on a given night during the commission of a crime, to asking with whom they last slept or what book they read, or what movie they watched—and in an effort to prove innocence, no matter the rights they might possess, many will answer the question no matter how inappropriate or damaging it might be to them. And it doesn't have to necessarily apply to criminals and government police states; it can be used as well between individuals and organizations. What prevents an organization from asking people their political preferences? Yes, these questions do run the risk of litigation but haven't stopped a lot of companies from adopting lie detectors, interrogations, and other highly invasive procedures.

The act of accusation, no matter the accuracy of a personal attack is more than sufficient to solicit a coerced response in an attempt to undo the damage caused by one's accuser; there are now untold cases of employees being accused of theft or sexual harassment by those with ulterior motives. Once the allegations are aired publicly, there is nothing to counter the perception now held by one's coworkers, friends, and family. With such loaded questions and pre-suppositional statements, the burden of proof is shifted to the accused rather than the accuser, an idea that sits afoul of the constitutional and legal requirement that the accuser prove guilt.

The same argument of revealing ourselves in the absence of guilt extends to a multitude of technological advances such as biometrics and drug testing; with facial recognition now found in many airports, the general premise is that if you have nothing to hide there should be no issue with getting photographed. It also implies a belief in a fair and just process of investigation. It is assumed that, by revealing the information sought be the questioner, that these revelations will support the claim of innocence; but it further assumes, incorrectly, that the process will be impartial and fair.

Privacy represents so much more than merely the right to be left alone, to find sanctuary from the relentless pace of commerce, technology, gossip, and all of the annoyances and invasions that pound incessantly at the door. And by now, I am sure you have become aware of how complicated this really is—privacy is not just about blocking telemarketers, it is an indicator of social health or lack thereof.

2

The Social Dimensions of Privacy

2.1 A Crash Course in Human Nature

"All legislation, all government, all society is founded upon the principle of mutual concession, politeness, comity, courtesy; upon these everything is based...Let him who elevates himself above humanity, above its weaknesses, its infirmities, its wants, its necessities, say, if he pleases, I will never compromise; but let no one who is not above the frailties of our common nature disdain compromises."

—**Henry Clay (1777–1852)**

Although a discussion of social psychology may appear to be an awkward companion to the disparate chapters devoted to biometrics and government surveillance and you may be inclined to question what relevance the so-called soft sciences have to do with privacy, the fact is to understand privacy requires a prerequisite understanding of the human condition that define the underlying attitudes towards privacy and freedom. That does not undermine the pivotal role of science but it is the decisions of people and not computers that define the boundaries of privacy and the technology is really little more than a tool to either enhance or diminish a person's privacy.

For the past two years I have had to wade through countless studies and volumes of research related to privacy in addition to many books on the topic and what troubled me was the obsession on "gadgets"—a single minded focus that did not do justice to the underlying psychological and sociological forces that result in invasions of privacy and violations of basic civil liberties. Chapter Two will hopefully accomplish several things, not the least of which is my desire to explain the underlying social mechanisms behind the technology and, more importantly, stress the point that privacy is an inherently human phenomenon.

What makes these dimensions of privacy so exciting is the infinite range of permutations that we can find in human nature; generalizations notwithstanding, for every theory there are exceptions that are unique to a person's experiences, beliefs, and environmental influences—no two people can ever see a painting the same way and that uniqueness makes the study both frustrating and enthralling because it serves to challenge common misconceptions and demands a fresh look at your own beliefs. This chapter isn't so much about privacy—it is all of the underlying human elements that influence it. Though by no means a treatise on social psychology I've covered what are believed to be some of the most important elements such as interactions, secrecy, gossip, mating theory, economics, and law to provide a quick "snap shot" of the forces that influence attitudes at an individual and group level.

While a majority of readers intuitively know what trust is and why they gossip amongst friends, seeing trust from a scientific point of view and reading about gossip from an anthropological perspective just might give you a new look on the world. These fundamental constructs act as a precursor to the discussion on organizations—after all, how can organizations that are comprised of people have respect for others if the people within those walls do not respect others. An organization is the aggregate of its members, and how they behave partially defines organizational culture.

Another central theme is the driving force of evolution and a built-in Darwinism that reminds each of us that 1) humans are a biological reproductive mammal, 2) there exists a natural desire for freedom from control, and most importantly 3) people will do whatever is necessary to ensure their survival. Ironically, the advancement of our cognitive capabilities coupled with the progress of civilization and the need for emotional bonding has blinded society to the often brutal realties of evolution and survival.

On a larger scale the same principles of Darwinism apply equally to organizations and their behavior frequently reflects a zero-sum mentality in the treatment of competitors, employees, and customers. And it is this 'all or nothing' culture that makes it difficult to believe a bank cares about your personal retirement plans or that your call is really being monitored for quality purposes, and so on, and so on. Mind you, if organizations were truly benevolent, there would be no need for unions or labor laws nor would we be subjected to the litany of Enron's or Tyco's. Why do you think there are Child Labor Laws? Remember that in many parts of the world children as young as five years old toil in mines because there are no controls or restrictions, and no appreciation for quality of life or respect for the individual. I'll come back to this question of evolution a little bit later in the chapter but in the mean time ask yourself whether you feel organizations are "evil" or whether they are simply trying to survive in an increasingly competitive and hostile market.

Nonetheless, organizations are comprised of individuals and it is the dynamics between two or more people that define relationships. What defines friendship, regardless of the type, between people? What attributes and persona's appeal to us and what criteria and profiles are employed to determine who constitutes a friend and who doesn't? Every one of us, all six odd billion people, apply a set of filters that define that kinds of people we find appealing and this is the genesis of trust and the willingness to share information with those around us.

Following the initial contact between two people, there is a relatively slow and often delicate process of information disclosure that stems from a natural curiosity to learn more about someone and over time, as the emotional bonds become more entrenched, trust is gradually established—don't underestimate the its value for trust is a key factor that determines the depth and quality of the relationship; a relationship without trust isn't a relationship—it is a causal acquaintance at best. There is, mind you, more to trust than just friendship—trust is required for the proper functioning of any culture and the absence of trust has frequently led to social isolation, conflict, and in some cases warfare. Trust, a simple short word, really defines our willingness to be vulnerable and exposed to others that we believe will act kindly and in our best interests. The same principals apply to people, organizations, cultures, countries, and geo-politics.

Closely related, there are some other concepts such as gossip and secrecy that will be covered in this chapter and which have a tremendous impact on what is or is not disclosed. Tabloids aside, it has been studied and argued that gossip is not only a force for good but necessary for the survival of human beings; gossip serves as a communications channel that not only informs us of our social environment but further serves as a means of behavior control and a measure of our position in the social hierarchy. Paradoxically the evolution of gossip conversely requires the parallel evolution of secrecy as a means of controlling both the disclosure and dissemination of private information to external parties. Throughout the course of history, secrecy has played a pivotal role and remains a key strategic defensive strategy that is utilized in the most casual of conversations to the murkier world of espionage.

All of these ideas—trust, intimacy, secrecy, gossip, and so on—are intertwined and each, when observed both independently and collectively, have a substantial influence on attitudes and policies towards privacy and these forces will help towards a better understanding of the shifting boundaries between what is considered private and what is considered appropriate for public consumption.

As much as person to person interactions are important to the study of privacy and freedom, few if any of us live in a vacuum and dealing with organizations as an employee or consumer is part and parcel of the daily "grind" and since organizations are defined as a collection of individuals with a single mission, all of the things I just mentioned are magnified exponentially. Organizations, whatever their charter, be it public or private sector, possess their own objectives and mission and the individuals in each respective organization also possess *their* own objectives and thus the forces of cooperation and competition can be felt internally within an organization just as they do externally within any given market.

Where privacy faces the greatest threat is when the power of a collective body meets the hostile landscape of the marketplace and where economics becomes the dominant player. The pressures of competition and economics alter the balance of power between organizations and individuals where privacy and respect are frequently marginalized if not discarded in pursuit of productivity and profitability; and these forces are in turn changing collective attitudes towards trust, one of the key forces that provides the glue between people and prevent social isolation and conflict.

However, nothing is to be gained by assaulting free market economics and time would be better utilized by thinking about how organizations and individuals can co-exist within a legally binding framework that clearly define the boundaries of privacy and personal freedoms. Somewhere along the polarized divisions of opinion exists a happy medium where individuals can respect the need for organizations to survive and, conversely, organizations can learn to respect, and even benefit from, the boundaries of personal privacy—the rights of one end where the rights of the other begin. Society is not at this point yet, not even close to it and as a result laws and regulations are required to ensure that a standard of behavior is enforced.

In thinking back to your childhood you will probably recall that your house had rules and those rules governed your behavior—violation of the rules resulted in punishments and adherence to same resulted in benefits. The same can be said of the growing trend towards privacy laws throughout the world and readers are offered a brief introduction to the laws that govern privacy from the United

Nations to the United States. Sadly, for every organization that understands and actively applies good policy, there will be just as many if not more that simply don't or won't want to understand that there are penalties for crossing the line.

While society is already over-governed, there is a need for rules to protect us from harm and the sheer volume of intrusions of one's privacy and violations of our constitutional liberties provides sufficient evidence that emphasizes the need for the rule of law; without the constitutional protection to express ourselves, governments could thwart the right to change governments leading to a police state; without the freedom to practice religion, many religious sects would be openly battling in our streets or controlled by the state—the protections most often ignored are those needed most. On a larger global, social scale there are myriad laws, statutes, and constitutions that simultaneously attempt to provide the infrastructure for good behavior while providing fundamental protections for the individual.

Privacy law is not exactly on the short list of your Friday night entertainment and the confusing patchwork of laws at the national and local level can be very confusing indeed. Still, a basic understanding of how privacy laws work illustrate the different approaches one finds in Canada, the United State, and the European Union. Beginning with a brief examination of the 1948 United Nations charter on human rights, the chapter also looks at the key legal frameworks in Canada, the United States, and Europe in an effort to provide a comparison of what privacy protections are offered in these countries in addition to the mesh of constitutional provisions and national privacy frameworks.

Chapter Two provides the underlying foundation for everything else that is discussed in this book and will hopefully provide just enough theory of behavior to get your head around why people invade privacy and what better place to begin than the interactions that occur between two people. Humans possess the need to communicate and this is one of the driving forces that bring people together.

2.2 Interactions

"Be courteous to all, but intimate with few, and let those few be well tried before you give them your confidence. True friendship is a plant of slow

growth, and must undergo and withstand the shocks of adversity before it is entitled to the appellation."

—George Washington (1732–1799)

There are, as you would expect, myriad factors and variables that affect your perceptions of the world and much of that owes to an individual's upbringing, beliefs, feelings towards others, and the persona projected to the "outside" world—introvert or extrovert, optimistic or pessimist, and the list of permutations is virtually infinite and it is even possible to exhibit dual attributes—in my case I am normally an introvert but can switch modes based on who I am with. The point being people are social animals and, unless they live under a rock, must co-exist with others.

And when an individual comes into (close) contact with another person, the interaction spawns a process of filtering and evaluation to determine whether the person in question exhibits a set of typical qualities against criteria that is determined by each of us—you may not even be aware that the process occurs but it does.

Of the more common characteristics, sensitivity, empathy, kindness, patience, compassion, warmth, and character are often cited as the key criteria on which we are judged and yet the primary driver tends to be integrity and honesty—it is these two attributes that ultimately determine the depth of the relationship and the degree of trust between two people. This set of filters and the process by which people are 'weeded out" is universal and generally applied during the initial phases of the relationship. Of course the measure of a person's integrity plays a role in determining the degree to which they are trusted and the extent to which privacy can be relaxed in their presence.

When speaking of a romantic relationship, regardless of the intensity or pace in which it develops, the same attributes are used initially and then coupled with sexual attractors that are based on the preferential profiles used during mating selection. It is worth noting, with some amusement, that the results of several studies including Olson and DeFrain 1977, shatter some of the stereotypical assumptions about sexuality with respect to the perceptions commonly held about the opposite sex. It has little or no bearing on privacy except as a reminder that we tend to overestimate the ability to "read" the opposite sex.

Men, as if it comes as any surprise, do indeed tend to focus on specific ana-tomical features during the mate selection process but the features that one would expect to rank highest are actually lower than you would expect in a man's prior-itization of features. While overall figure and sexuality ranked first and second respectively, the "features" that you would think to be a man's first priority actu-ally come in *tenth* place—yes tenth—that doesn't mean men don't look, but it isn't always the first thing that comes to mind. While running the risk of being pilloried for my thinking, and perhaps it is a reflection of the political or cultural climate, but there is this habit of conveniently forgetting that humans, and by extension men, are *reproductive* organisms hence to not "look" would run counter to millions of years biological development.

Women, quite surprisingly, ranked general attractiveness and sexuality in *first and second place* with sensitivity and build coming in a rather distant seventh and ninth respectively. From this it is quite evident that men and women really do consider sexuality equally significant but with a slightly different focus: men focused on a women's figure while women took a more "holistic" view in consid-ering the entire person. Ironically, character ranked much lower than would be expected but remember that two different scales are being applied during the selection process—the first filters based on physical attraction and the second fil-ters based on personality. Unfortunately the politics of sex has blurred and to some extent polarized opinions despite overwhelming evidence to the con-trary—there is much more in common between men and women then some would have you believe.

There is actually a highly structured order in which adults apply these pre-defined filters that effectively narrows the pool of possible friends or partners to a few prospective candidates—obviously additional metrics apply to partners as opposed to friends but there is a lot of common ground in the process (Kerckhoff and Davis, 1962). The first and obvious filter is an eligibility and propinquity fil-ter which limits the pool to those that are within a relatively small geographic area and that we would come in contact with such as those we work with, go to school or church with or engage in activities that result in social interaction. To put it another, your true "soul mate" might live in Europe but if you live in New York, chances are you will never meet them due to the logistics and the tendency to pick from those we already know or are within a reasonable range. These criteri-ons make for an interesting observation—given the likelihood that a person will find someone within a given geographic area, and barring random chance encounters beyond this boundary, it stands then that a person's so-called soul-mate, if you accept the premise of such, is often right under our nose.

The second filter acts to eliminate those candidates that are not in the same social, economic, age, or cultural groups. As a general rule people will *not* stray far from their peer and cultural groups and this is evidenced in interracial marriage statistics that indicate, in one example, that less than 7% of all marriages are between Caucasian and Asian cultures while Caucasian and other races varied considerably. Despite the mantra of cultural diversity and political correctness cross-cultural relationships are not yet the social norm. Here too if a so-called perfect mate (at least physically) does not reasonably fit within a person's socio-economic class then it is highly possible they will be eliminated from the pool. Similarly, you could meet the perfect mate but time could conspire with or against you depending on where each person is in his or her live; time also plays a significant role in the random interactions in which two individuals may or may not cross paths.

Insofar as physical attraction filters are concerned, from the beginning of sexual activity, adults adopt a *profile* of preferred attributes that determine the type of person they seek and there are a number of stereotypes that have persisted over the past 50 years such as the mantra that blondes have (or are) more fun. Consider the fact that less than 7% of women colored their hair blonde before 1955 but following the landmark advertising campaign of Foote, Cone, & Belding that number soared to 413% in a single year. Furthermore, according to a study in the 1985 article in the *Journal of Social Behavior and Personality* volume 13 entitled "The Influence of Hair Color on Eliciting Help: Do Blondes Have More Fun?", it was shown that women helped men or women equally *regardless* while men were more likely to help a woman than a man—but the color of hair had no bearing whatsoever. On a more social level, The *South Wales Argus Study* found a brunette subject received *more* attention from men than she did donning a blonde wig. As the statistical studies indicate, it is difficult to determine why we prefer specific attributes but they clearly influence the selection process. It is also important to keep in mind that these are generalizations, which means that a majority of respondents responded in a particular pattern but it does mean there aren't exceptions.

As a mental exercise, you can try to apply the filters that have been covered so far by thinking of three or four people that you find very attractive and attempt to determine the common attributes—chances are you will find there are "elements" of each that occur in each of the people you select—the same process works in reverse by first defining the attributes you find attractive and then identify three or four that fit the profile—it doesn't matter who, it is just an exercise. While my own so-called "favorite four" shall remain conspicuously absent there are a few more filters to consider.

This filtering process continues with two more filters: compatibility and balance. It stands to reason that however attractive a partner may be, there is still the risk that neither of you will share compatible attitudes and interests that will sustain the relationship over a *long* period of time. These filters actually work on friendships as well—two friends need to have compatible interests and personalities and both need to feel the friendship is appreciated. If or when that relationship transcends the boundary of friendship there are further evaluations of basic religious and political beliefs, attitudes towards money, work or people, behaviors towards division of labor and balance of power, gender role expectations, and personal preferences such as leisure sports that will determine the degree of compatibility.

Finally there is, even at a subconscious level, an assessment of whether there is a balance in the relationship. Are you bringing more into the relationship than the other person—this need not be a perfect balance, and it will vary between two people, but it is assumed there will be a degree of *equilibrium* in a relationship. This is just the beginning of any friendship or romance and requires a lot of searching and probing to get the answers required to effectively filter those that are not compatible. Even just as friends or co-workers there are similar evaluations of balance: does your colleague contribute as much as you do? Do you find you are calling a friend far more often than they ever call you? If or when these inequities are discovered or perceived there is usually a "re-assessment" of the relationship.

All of these factors play a role in privacy as a relationship continues to develop and as friends or partners seek to learn more about the other—obviously we can not make informed decisions about a friendship or relationship without knowing who is sitting across from us. Curiosity is a pivotal player in this game and it's worth discussing in more detail.

While our contemporary culture would appear to disdain personal questions, more often than not the questions stem from our curiosity—and it cannot be stressed enough that curiosity isn't just healthy it's necessary for the survival of this species and you are only doing when you were genetically programmed to do—to seek and discover. And this natural desire to learn more often results in occasions in which a friend asks a very, very personal question or perhaps you have inadvertently asked something that was "over the top"—and quite often the question is immediately followed by an apology for presumably overstepping the boundary of a person's privacy; but there remains this insatiable desire to understand the other person, particularly if they are a very close friend or someone for which there is a strong attraction.

Curiosity is an unusual attribute. On the one hand it is celebrated and encouraged in children and yet often discouraged in adults. Children are taught to discover their environment and learn but as adults, particularly in the workplace, curiosity is replaced by risk mitigation where the overriding concern is staying within the boundaries. It might even be argued that those that frequently flout the boundaries are also those prone to greater discoveries because they are not constrained by the same mentality; witness many of the great discoveries in science where the explorer is often reclusive, eccentric, and not the most socially adept amongst his or her peers.

Several studies including the Kashdan & Rose Study at the University of Buffalo found that the degree of curiosity influences personal growth, happiness, opportunities and the level of intimacy in relationships. In this context, curiosity is viewed as a positive emotional-motivational system associated with other systems including recognition, learning, pursuit, and self-regulation. I would also add adaptation to this association since curiosity leads to discoveries that ultimately alter our worldview and by extension alter our behavior. Put simply, those with a higher propensity towards curiosity tend to seek more information and novel experiences which we might refer to as the *breadth* of their experiences but they are also more inclined to exhibit a higher state of flow in which they absorb more as well—that in turn leads to a greater *depth* of experience that also enhances the quality of life

Curiosity plays such a fundamental and pivotal role in the development of a person in particular and the overall evolution of the species that no discussion of relationships, trust, and privacy would be complete without it. If you accept the

Carl Rogers Theory of Personality which posits that the people are generally good natured and mean well coupled with the natural desire to learn then it becomes even more obvious that sensitive questions from friends or co-workers are not necessarily meant to cause harm but rather to learn more about the person behind the persona. The greater the trust the more intimate the communications; the idea of depth and breadth are revisited since they have a bearing on how deep a friendship can go based on the limitations or wishes of one or more parties.

Not surprisingly, everyone maintains a spectrum of friendships that covers the spectrum from co-workers and casual acquaintances to deep, life long friends with the number of the latter far smaller than the number of the former. As one traverses this continuum of intimacy, there is a corresponding increase in the level of trust that is inversely proportional to the levels of privacy that are required between two parties. Things that would never be appropriate to disclose with a casual co-workers may be welcomed in a deeper friendship. This has a direct bearing on one's expectations of privacy; the deeper the relation, the more intimate the disclosures. And the filtering process, coupled with the principles of trust, determine to a large extent who is and who is not allowed to cross the line.

This ultimately sets the boundaries of privacy by determining who is entitled to information and the assessment of risk of disclosure—the less likely that a person is assumed to disclose personal facts the more they will be entrusted with them. While time also plays a role in establishing the trust there are no hard and fast rules as when it occurs—it is still a largely subjective decision process. All of these principles, from the selection of mates and friends to curiosity and trust, have a significant bearing on the levels of privacy that a person requires—healthy, deep friendships or relationships negate the need for excessive privacy while shallow, inconsequential relationships demand a more robust defensive perimeter.

2.3 Concepts in Trust

"For it is mutual trust, even more than mutual interest that holds human associations together. Our friends seldom profit us but they make us feel safe...Marriage is a scheme to accomplish exactly that same end."

—H. L. Mencken (1880–1956)

Trust has been studied and applied to virtually every major field of study including science, computing, economics, sociology, and law, and there is a large body of literature devoted to it and its applications. Having mentioned it several times, I want to revisit the subject in more depth. You do not need to be a Professor of Ethics to apply trust and it is something you have been doing intuitively since childhood. Still, a bit of theory is a good backdrop to how trust works in the context of privacy.

According to Rousseau, trust is the psychological state comprising the intention to accept vulnerability based upon the positive expectation of the intentions or actions of another person. It is this *interdependence* between people that requires a degree of trust to compensate for the additional risk assumed in any relational interaction—conversely, the more dependent we are on others, the more information we require to make these determinations—it's a bit of a Catch-22. Understandably, the propensity for individuals to trust other people varies widely from one person to another and evidence clearly shows that trust behaves more like a continuum than a discrete value; it is rarely the case of declaring *absolute* trust or distrust but a broad continuum of trust from none to absolute nor is it applied equally. It is further based on a set of expectations that a person will cause no harm and are capable of doing what is promised. Trust is often based on several factors including *perceived* ability, integrity, and benevolence which taken together forms the level of trust they are granted. While ability is clearly understood, benevolence is the assessment of the trustor that the trustee will not only cause no harm but will act to further our interests or, at least, not interfere with them. Think of it as a type of contract or alliance.

Another formal construct is the *level* of trust that is maintained between parties and the decisions that determine the assignment of levels. During the initial stages of any social interaction, regardless of the context of the relationship, there is a tendency to utilize a calculus-centric approach that is a cognitive driven process in which one person calculates how another person is likely to behave in a given situation and whether that person will behave *consistently*. In addition to this assessment there is a co-existing reward-punishment balance at play whereby the *trustor* communicates to the *trustee* unspoken and understood implicitly by both parties; if a secret is shared with a friend, the consequences of violating that trust are appreciated by both sides. But for this to work effectively, both parties must share universally accepted beliefs about what constitutes acceptable conduct and this may vary in some cultures—failure to recognize a common ground of accepted conduct means the relation is bound to result in a very short lived relationship. For example, consider the code of silence that has always been the hallmark of organized crime—to violate such a trust invites a less than pleasant swim with the fishes and all members of the organization understand this even if it is never spoken. Informants in a prison are equally quick to discover that there are consequences for their actions, and historically those that have been accused rank no better than sexual predators.

As a simple example, the varying attitudes and behaviors towards voicemail and e-mail etiquette illustrate what happens when two parties follow different standards or hold different expectations. If one person has the expectation that calls ought to be returned in a reasonable period of time and the other never returns e-mails or calls or does so in a way that suggests a lack of respect concern than a conflict is going to eventually emerge. If the trustee does not communicate to the trustor that there is an issue than the trustor may be left to draw his own conclusions even if those conclusions are presumptive.

This concept of reward and punishment is critical even though it is rarely explicitly set forth. If a secret is shared between two people and the former states "You can't tell anyone" then it is assumed by the former that the listener understands the gravity of the disclosure and the cost of disclosing the secret and there is the expectation that the listener will be consistent in not revealing secrets—in the absence of consistency there is not trust, and with no trust there can be no relationship.

While the calculation of integrity plays out at home and amongst friends the dynamics can be quite different in the workplace where the friendships between co-workers are more "fragile". If someone is known to lie or cheat or take credit for the work of others then this person likely demonstrated an *historical* pattern that is known to his or her peers; if someone else is known to be manipulative, deceptive, or lack concern for others then he or she too has demonstrated a pattern of behavior and eventually these patterns become widely known. I am somewhat divided as to whether these people recognize their behavior and the damage they cause or whether they are blinded by their own agenda. It remains an open question.

It has been theorized that this calculus *shifts* as the relationship becomes deep and entrenched where the trust is referred to as an *identification based trust* where both parties are able to identify a set of shared goals, beliefs, and values that are important to both parties. There is speculation that this, the highest level, results in an *internalization* of desires and intentions—each understands what the other cares about and *selflessly* act to protect and enhance their respective needs. Don't underestimate the strength of the emotional bonds; it posits an emotional dependency between two people that make betrayal unthinkable and the loss of the friend unbearable—but the friendship is so deep at this level that betrayal doesn't even enter into the equation and if it does the friendship never reached this level. I am not entirely certain whether everyone has been so lucky to have found friends of such depth—this too is an open question.

How fast does a trust develop between people? Obviously, trust will be granted over a period of time and the level will be increased as the relationship matures; but how fast or slowly do people develop a sense of attachment? The speed at which this occurs is dependent upon the individual and circumstances will differ for everyone but there is a lot of variance—some may take years to unfold while others are virtually immediate. It does seem possible to achieve trust very quickly but this increases the risk that it was misplaced due to lack of diligence or clouded judgment. For all the talk of theories it is still our instincts that guide these decisions—sometimes we get it right, and sometimes we don't.

But what happens when a trust is betrayed? Everyone, at some point, is going to be betrayed, it is almost a right of passage. A betrayal, so there is no confusion, occurs when the trustor's positive expectations of the trustee have been violated. At this point the trust is either lowered or destroyed depending on the severity of the violation. There are different types of violation and different degrees of severity that affect the trust between two people.

Although it can be rather unfair, a failure to live up to one's expectation can diminish the expectation that we will live up to others. If for example, a person is late for a drink or dinner it may not diminish the expectation of reliability the first time but if was to happen on several occasions that it is reasonable to expect the trustor will have a diminished expectation—it is assumed thereafter that this person will be late. Does erratic or disrespectful behavior in one context become associated with other expectations? I believe that there is a direct connection between the two and systemic patterns of behavior would have a bearing on unrelated decisions on whether to trust that individual.

In the case of a severe violation of trust, where a secret is released or a violation of equivalent damage, there is a lot of research to suggest that two parties *can* repair a damaged trust; however, it is difficult to image that many friendships will survive that kind of betrayal or whether you could trust them again—the real question is whether, in light of this knowledge, can a friendship really return to its original form—it may be possible to forgive but to forgetting a transgression is something else. In the deepest of relationships it would require a re-commitment to the offended person, and a genuine desire to remain as friends. I am personally divided on this question of violations; though somewhat harsh I tend towards the school of thought that holds that friendship is terminated at the moment a *severe* violation of integrity occurs.

It depends in large part on an aggregation of factors including the nature of the offence, the severity of the offence, the intent of the trustee, and the lengths to which the trustee has gone to repair the damage, and it does not require that judgment will be even handed—if a Person 'A' was to reveal a personal secret and Person 'B' had totaled yours, which offence is *greater*? It might be tempting to immediately assume that the latter is greater than the former, but what if Person 'B' was involved in an honest accident, showed profound remorse, and took immediate remedial action to compensate you for the damage? What if Person 'A' was intoxicated or coerced? What if Person 'A' was equally sorry for his or her transgression? Think back to the last time you were betrayed and apply the same metrics—would it soften or harden your opinion of either party?

The trust between individuals and organizations, if it truly exists, embodies many of these attributes, but they are always based on the calculus approach. If a website promises not to disclose or sell your personal information to marketing companies, you must make a decision whether you trust them to keep their word based on their history, public image, and reputation in the marketplace—you are effectively calculating the cost of disclosure and benefit based on these factors. If, as it has already happened, a site sells its customer lists due to financial hardships, or routinely provides information to third parties in violation of its privacy policy, then trust in that institution will be diminished. But the impact of that betrayal has much broader consequences if the public decides, rightly or wrongly, there are systemic abuses with *all* institutions. In short, it hurts the legitimate players in the market when even one bad player abuses the trust of a consumer. It is really no wonder that banks and many organizations go to such great lengths to create an image of trust and reliability and it is fair to question how much of it is just smoke and mirrors—the time and resources would be better spent actually delivering on the promises that have been made.

Non-Associative Trust

The principle of Non-Associative Trusts means that trusts are not linked. In computing terms, it means a user is trusted to log into the web server but the trust does not necessarily extend the privilege to login into the mail server—they are not mutually inclusive. The same principle can be extended to relationships and there are no "absolute" values—it is a constantly shifting dynamic where certain people will be trusted with some things but not others and the equation changes constantly according to circumstances; the accountant can be trusted to balance the book but that trust does not automatically extend to watching the house or babysitting the kids.

Non-Declarative Trust

A trust cannot be asserted or solicited in the form of "You can trust me" or "Don't you trust me?". You cannot coerce someone to trust you and if nothing else it will have the opposite effect. However, the statement "I trust you" is valid if it is unsolicited. The trust placed in an organization is not the kind of trust the organization was hoping for—it is at best fleeting and subject to change based on the perceptions of a consumer or employee. Hence the advertising push to trust a bank probably won't achieve its objectives.

Asymmetric Trust and the Balance of Power

Of all the forms of trust one of the most interesting is this idea of a "Balance of Power" because it illustrates the delicate negotiations between two people as they seek to learn more about each other. This concept, though active in all relationships, is particularly remarkable in the course of courtship where disclosures tend to be much more intimate and therefore more risky.

First, I need to explain this idea of Asymmetric Trust. In a nutshell, trust is not always equal between two parties; and the idea of Asymmetric Trust holds that just because Subject 'A' trusts Subject 'B', there is no obligation to reciprocate. While this might hold true for a very brief period of time in very casual relationships, I would hypothesize that this asymmetry runs contrary to the requirements of a deep interpersonal relationship, so I will introduce a new concept referred to as *Trust Equilibrium* where both parties exchange information equally at a rate that does not leave one party overtly vulnerable. If Subject 'A' continually discloses vulnerable personal facts and Subject 'B' receives but does not reciprocate, there is an *unhealthy* shift in the balance of power. At some point Subject 'A' will, it is assumed, recognize the imbalance and take corrective actions that might include silence, ambiguity, and other means of reducing the risk of further disclosure, and this may cause deterioration in the level of trust and eventually in the friendship itself.

Non-Distributive, Non-Transitive Trust

Finally, you cannot arbitrarily share or pass along a trust to someone else. If Subject 'A' trusts Subject 'B' and Subject 'B' trusts Subject 'C', it does not logically infer that Subject 'A' can trust Subject 'C'. That might seem a bit confusing but if you think of several friends it becomes self-evident. Anne trusts Mary and tells her what a terrible day she is having at work. Mary trusts Bob and tells him about the terrible time friend Anne is having. Now Bob, an *un*-trusted person, knows something personal about Anne and which Anne did not consent to. No matter how much Bob is trusted by Mary she has still violated the trust of Anne.

As a fitting closure to this segment, trust is a fundamental prerequisite to every form of human interaction from friendships to geopolitical alliances and at the heart of trust is the expectation that the other party will be a benevolent companion. And the extent to which a person comfortable with another is set, at least partially, by the expectation that they have their best interests at heart and at the very least will cause no harm. Though I have mentioned this earlier, it does not hurt to reinforce its importance. Trust is all the more challenging when the circumstances are undefined and ambiguous—its one thing to trust the babysitter to watch over your child because its assumed there is a history upon which that trust is based—I'm not sure you'd feel the same if you had to count on this babysitter to rescue you from a burning building—there is no *historical* evidence in your relationship with that person to determine how he/she will react under adverse circumstances. And in the absence of certainty, you'll be forced to rely on this person's benevolence—does he or she have what it takes to put your interests above their own or at the very least not interfere with or cause harm to your interests.

2.4 Gossip, Secrecy, and the Art of Information Control

"To harken to evil conversation is the road to wickedness...(Pravis Assuescere Sermonibus Est Via Ad Rem Ipsam)"

—Anonymous

Despite protestations to the contrary, gossip has been a favorite pastime since the dawn of civilization and that isn't going to change. The relationship of gossip to privacy stems from the need for information control mechanisms in a person's personal and professional lives and it defines, in part, how information is both disclosed to external sources, and more importantly how the information is distributed once it is released. It is safe to assume that once a material fact is released into the public domain it can never truly be retracted so a discussion about gossip is well within reason and the management of rumor and innuendo should be part of your own privacy strategy.

From a historical perspective, gossip has been a basic communications vehicle of the human species since the genesis of speech; from the Stone Age onward, it served numerous functions including the dissemination of critical information such as the location of food, shelter, and other threat assessments. But it also served a multitude of other purposes just as it does now. In seeking the formal definition of the term, you can look to any dictionary and find entries for gossip:

- *Rumor or talk of a personal, sensational, or intimate nature*

- *A person who habitually spreads intimate or private rumors or facts*

- *Trivial chatty talk or writing*

There are other definitions but they characterize gossip in the same manner even though it does not do justice to the immeasurable social importance of gossip or explain why society requires gossip for its survival nor does it explain the reasons that drive us to do so in the first place.

First of all, recalling the importance of curiosity and its impact of societal and psychological development, gossip can be viewed as a form of social comparison because people want to know what people are doing or saying as means of finding commonalities. It is much easier to withstand the loss of a companion or a job with the knowledge that others are going through the same experience and it serves to put a problem or crisis into perspective. And in some schools of thought, gossip, given these comparisons, actually results in greater social cohesion, though it does not satisfactorily address the negative dimensions and damage that is caused.

Gossip occupies a higher pedestal in social development than most would otherwise appreciate and it fulfils a set of social functions above and beyond the proverbial dishing of dirt. In passing gossip about a co-worker that dresses poorly or is rumored to be seeing someone, is it really a case of passing along a rumor, or is it to send *signals* to that individual that they are stepping over *accepted boundaries of behavior*; the short answer is that both are true. It not only allows a group of people to learn about others in the same social group (or even a different social group) but it also works as a means of comprehending the limits of behavior that are permitted in a given culture. The message conveyed to all is: follow the norms of accepted behavior or find oneself at the mercy of the whispered word. Those that have crossed the line will usually attempt to clean up their act and those receiving gossip will know where the line has been drawn. So despite its surface pettiness, gossip is a form of collective behavioral control.

Using the workplace as an appropriate metaphor, every office of every size follows the same pattern and there are different communications channels within every organization—reading the company employee manual or sitting through yet another positive spin speech will provide the official story but rarely if ever does it capture the undercurrent of communications. Assuming a key employee was leaving an organization it may be stated quite safely that some or most of the employees know about it long before it is announced and efforts to control or quash covert channels are futile at best.

Like the top secret "flash" traffic between embassies, it is the grapevine that carries the most important traffic such as who's in, who's out, who's important or not and whatever the dirt of the day might be. Pay closer attention to those who are transmitting and those who are receiving—it isn't always the case of what is being said but rather by whom and to whom it is being said. The "Receivers" within any social institution may not be the seniors of the organization but they generally receive all of the secrets that are floating around—just don't expect to get anything out of them, their strength lies in their silence.

Likewise those that are frequently the subject of chatter are deemed more important than those that aren't and society determines one's status by the extent and frequency a person receives; any read of the National Enquirer will tell you who is hot and who is not because of what is said but equally by how often it is said. As superficial as all of this appears, studies have shown that gossip can be a potentially positive force and that less than 5% of what we hear is negative. On the bright side, it permits an understanding of people and deals with common fears, anxieties, and other maladjustments that occur in most adults at some point in their lives. On a larger scale, it has the ability to challenge political power, maintain socially acceptable behaviors and overturn negative institutions.

It is extraordinarily tempting to assume that small people engage in small talk while the intellectual elite engages in ideas—an assumption made all the easier by the proliferation of tabloids and reality television that caters to the lowest common denominators; however there is little if any evidence to support this belief and IQ is not a determinant for gossip and it is not unreasonable to expect that even the greatest minds, that top 5% of the worlds population, engage in idle gossip just as much as the other 95%—for all of their intellectual capacity they are no less vulnerable to the trappings and temptations of human nature.

Gossip is often the product of uncertainty. This uncertainty exists in the absence of fact and results in the genesis of rumor and speculation that have no basis in fact but take on a life of their own. It could even be argued that closed doors and quiet meetings are actually more damaging than the truth for in the absence of facts any population will be inclined to come to their *own* truth. And that brings me to the negative impact of gossip—not only for the target but the purveyor of information and it highlights the maxim that things are rarely as they may seem—at the very least it ought to make you rethink exactly what is you're hearing and the confidence you have in both the provider and the information.

There is always the risk that there will be those that bring a malicious intent into the equation and seek to damage the target for whatever reason or perhaps they are simply malicious by nature. There is also the risk that the rumors that circulate are the product of human error or exaggeration. Even if the rumor was accurate and true, there is the potential that those providing gossip will find themselves the victims of the "Boomerang Effect" where the provider becomes the new target because of their treatment of the intended target. Referred to as

Spontaneous Trait Transference, the person that is listening will tend to associate behavioral traits with the speaker. If a politician complains of dishonesty he will be viewed as dishonest; if a lady in your office complains of someone else's affair, she is viewed worse than those engaged in the affair. And both of them are view as untrustworthy. What a person says of others is even more a reflection of them than it is the intended target.

Finally, before discussing the opposite side of the coin, secrecy, the ability to think rationally affects the value associated with the information that is received and applying this to gossip can be quite educational. Intelligence Services around the world do two things: *they collect information and they evaluate it*—and it wouldn't hurt to adopt similar practices.

What is the *source* of the information that is received? *Who* provided the information? Is it *accurate*? What are their *motives*? How do you know the "target" wasn't the originator of the rumor? Is the rumor really true or did the perceived target engage in misdirection and plant false information with the knowledge it would be leaked. If the CIA can use tactics like this in the protection of national security then there is reason to expect they would work under different circumstances.

There are hundred if not thousands of techniques and variations but the use of misinformation is timeless and effective. Here are three examples

- *CIA tells a suspected double agent of a fictional air base. If there is increased surveillance or the CIA hears of it through another channel then there is a good chance the spy talked.*

- *Johnny tells Martin that he likes Lisa—but he doesn't, he likes Alice; he just wants to see how Alice will react.*

- *Bob is a genius of a software developer but not appreciated—so he lets the rumor out just before his performance review that he is considering his options just to put additional pressure on management or to see what they really think of him.*

Well no one is encouraging you to engage in subterfuge but it does go to illustrate that what you hear isn't necessarily true—it could be the truth or it could be what someone *wanted* you to hear and that emphasizes the need to think carefully of what you are hearing and the source of the information.

The Art and History of Secrecy

Just as there is a human and societal need for gossip there is a diametrically oppo-site need for secrecy in the hopes of preventing aspects of our lives or activities from either being disclosed or abused. Nowhere in the history of mankind is there a culture in which everyone is trusted with everything—there is a selection process whereby humans determine, based on trust and other criteria, who they can entrust and with what. Even liturgical and semantic analysis of the Bible pro-vides sufficient evidence that Jesus was a master not only of secrecy but plied the tradecraft with skill; contrary to the synoptic gospels, the truth of God was not revealed evenly or publicly bur rather through selected individuals, and even then never completely. Ironically, the words of Jesus were conveyed through gossip and the grapevine. The Bible is also littered, if you will pardon the term, with cryptic passages replete with double meanings, misstatements followed by clarifi-cations, and myriad deceptions and ambiguities. The Bible is just one of many sources of information that provide evidence of secrecy and there are a variety of tactics employed to ensure a secret remains secret, what is often referred to as Information Control.

To achieve this end game, different techniques including deception, lying, misdirection, ambiguity, and even humor will be used to ensure a secret remains safe. Governments in every country around the world, and the politicians that survive within those labyrinths, are the consummate masters and engage in the aforementioned practices that we would generally find repulsive in our own social circle. And yet, as I just illustrated, people engage in the same tactics albeit on a much smaller scale.

Secrecy is also used to define personal and organizational power structures by further creating a polarity between those in the know, the inner circle, and those that are left out of the loop. Power emanates from above and to frustrate matters, in most cases, it is not a matter of asking but rather of being told and constitutes a form of mediated information release to a select group. This form of mediated release occurs in organizations every day except now we are limited by non-dis-closure agreements.

The privacy issue associated with gossip and secrecy comes down to control-ling what information is going to be released and to whom it is going to be given. If an employee was considering a transfer to another division within a company the decision of whom to tell and when becomes critical: if spoken too soon or to the wrong person his/her ambitions might well be sabotaged by a jealous co-worker. What if the same employee was highly valued within a particular working group but wanted out—there is the possibility the supervisor might also sabotage that effort to ensure the person remains in the group even if such actions are short-sighted and terribly transparent. Better in this case to make the transfer and announce it *after* it is complete.

2.5 Economic & Legal Influences on Privacy

"The makers of our Constitution…sought to protect Americans in their beliefs, their thoughts, their emotions and their sensations. They conferred as against the Government, the right to be let alone—the most comprehensive of the rights of man and the right most valued by civilized men."

—Justice Louis D. Brandeis dissenting in the Olmstead decision

There is no point in belaboring the issue of organizational abuses of privacy beyond that which is absolutely necessary except to set forth the argument that there would be no need for protections against abuses if they did not exist in the first place. With that said, the willingness to invade the privacy of consumers or employees owes itself to the pressures of contemporary economic competition and in the face of such competition it becomes much easier to rationalize organizational behaviors that would be unacceptable in any other situation.

Then again it is not that difficult to rationalize any behavior even when these rationalizations are either transparently self-serving or act to justify an indefensible action. In particular, as a means of proving the fallacy of the typical spin, Fleet Financial Group laid off 3,000 employees despite a year of record profits while providing excessive bonuses to its executives. The management of Fleet argued that this was necessary to prevent even greater job losses in the future and the loss of key management personnel. All of which would be valid if and only if, the long-term threat actually existed.

Even then there is a school of thought that supports the morality of these actions even in the *absence* of a threat. Referred to as Conventionalist Game Ethic, the argument goes that if everyone pursues his or her own economic self-interest then society benefits as a whole. Thus, as it has been observed and speculated for years, we have entered into a period of Economic Darwinism, a no-rules, no-holds war in which only the fit shall survive and which Governor Angus King referred to as an unsentimental, unremorseful competition that individuals are not accustomed to. Herbert Spencer had coined the expression "Survival of the Fittest" in 1867 and it is synonymous with the principles of natural selection put forth by Charles Darwin. Although I will come back to this idea of evolution as a driving force of privacy and human rights, this era of Economic Darwinism does not bode well for privacy. An unsentimental economy has no patience for the inconveniences of civil liberties since such inconveniences are a *barrier* to achieving optimal operational efficiencies.

For that reason society requires a set of rules that define what is permissible and what is not and these rules have evolved albeit slowly to provide a basic set of fundamental protections that guard against systemic abuses of civil liberties including privacy and the law is mechanism we rely upon to define and enforce these protections. Unfortunately, the onslaught of sensational cases coupled with abuses within the legal professional has obscured, maybe even damaged, the meaning of law.

Of course prior to the contemporary legal system to which most have become used to, there were innumerable systems throughout the history of law that have preceded the current view, and they were comparatively brutal when viewed through contemporary standards of crime and punishment. Before looking to more modern approaches to justice it is worthwhile to take a (very) brief journey through the legal systems of centuries past. By no means a treatment of legal history, I only want to touch upon a few periods of law to illustrate the doctrine of crime and punishment.

The biblical expression "an eye for an eye" is as good a placed to start as any and, however barbaric it may seem now, dates back to the Babylonian Hammurabi's Code of the 1750's where it was taken quite literally. If a house collapsed because the builder did not make it strong enough, killing the owner, the builder was put to death. If the owner's son died, then the builder's son was executed. Similarly defamation of character by today's standards is quite tame compared to the punishment of losing your tongue and the punishment for cheating on one's spouse were, well, painful and rather permanent. That of course is only one code of many that have evolved over the centuries, and thankfully more humane systems have since taken root with each system generally more humane than others.

Again, I do not put this to you as a complete account of legal history but it does indicate that from the beginning of the earliest laws, a set of actions and consequences.

It further establishes that there are limits, defined by a set of rules that define what one person can do to another and what actions will be taken when those boundaries are violated. Virtually every system of law sets forth boundaries of behavior and the punishments for violation of those boundaries—The Book of Punishments, a legal book printed in China, set limitations on the ways in which someone could be punished, even if those punishments ranged from tattooing to amputation and death and the Chinese Code of Li k'vei (the first imperial Chinese code of law) established the standards and codification of behavior even though the punishments were often very grisly.

Everything we do, from the time we drive to work and make our way home, is in some way governed by a set of rules that are imposed by those in power in accordance with community standards of morality and social customs. And let there be no confusion, a nation's "laws", referring to the rules imposed by governments, are specifically drafted to alter and control human behavior, and in so doing they define what we can and cannot do with a prescribed penalty for non-compliance. If you speed, you get a ticket; if you commit a serious criminal offense you go to jail—and different countries treat offences based on their philosophical foundations—American justice tends to be overly harsh while Canadian justice tends to be incredibly lenient. At any point where people live within a structured social environment, from the smallest of villages to the largest cosmopolitan centers, laws have been required to provide the cohesive rules of conduct that prevent us from overt chaos and warfare. If people were allowed to choose at random which side of the street to drive on, driving would be dangerous and chaotic.

Laws that regulate business affairs help to ensure that people keep their promises. Laws against criminal conduct help to safeguard our personal property and our lives. Even in a well-ordered society, there are going to be disagreements and conflicts are going to arise and when a consensual resolution cannot be reached then society becomes dependent on the law to resolve these disputes peacefully—it doesn't mean the outcome will be to ones liking, but the basis of the law assumes fairness and equal treatment under the law. If two people claim to own the same piece of property, it is hoped that civilization has progressed to a point where such disputes are not resolved in a duel—instead the matter must be resolved through a more cultured, sophisticated approach that is predicated on logical argument and debate and let institutions like the courts decide who is the real owner and make sure that the real owner's rights are respected.

So it suffices then that the law is necessary for a peaceful society where rights are understood and respected—for example, laws ensure an element of fairness and that includes the right to privacy and all basic civil liberties. There is a further distinction between different branches of the law, such that public law governs social behavior and applies to all while private law focuses on the relationships and disputes that arise between conflicting parties.

Common law is also crucial since it is followed in every province except Quebec and every state in the United States with the exception of Louisiana and it can have a marked impact on privacy decisions. For the sake of reference, common law is a system or branch of law that originated in England and based on the medieval doctrine that the law administered by the king's courts represented the common custom of the realm, as opposed to the custom of local jurisdiction that was applied in local or manorial courts. In its early development common law was a product of several British courts including the King's Bench, Exchequer, and the Court of Common Pleas. The term "common law" also refers to the traditional, precedent-based element in the law of any common-law jurisdiction compared to other branches such as statutory or legislative law.

The distinctive feature of common law is that it represents the law of the courts as expressed in judicial decisions. The grounds for deciding cases are found in precedents provided by past decisions, as contrasted to the *civil law* system, which is based on statutes and prescribed texts. Aside from the system of judicial precedents, other characteristics of common law are trial by *jury* and the doctrine of the supremacy of the law.

Originally, the supremacy of the law meant that not even the king was above the law; today it means that acts of governmental agencies are subject to scrutiny in ordinary legal proceedings. It is loosely interpreted to mean that all men are equal before the law and that no special preference is given to wealth, rank, or position. Though noble in its intent, the recent cases like the O.J. Simpson trial fairly call this concept of supremacy into question—while the judge is presumably impartial, their appears to be nothing to prevent a defendant from "stacking the deck" in their favor. Contrarily, and to ward off the anticipated diatribes against the legal profession, we would expect the same right afforded to us if our freedom was at stake.

Judicial precedents derive their force from the doctrine of *stare decisis* [Latin translation to stand by the decided matter], such that the previous decisions of the highest court in the jurisdiction are binding on all other courts in the jurisdiction. However, changes in technology and social conditions render many decisions inapplicable except, as a form of analogy and court must then look to the judicial experience of previous case law. This ability to base decisions on the past while taking into consideration contemporary conditions provides the system with both stability and flexibility—the flexibility stems from the ability to deviate to accommodate social changes within a reasonable framework and the stability is gained through the authoritative decisions of the past.

Nevertheless, as many cases have shown, the courts have failed to keep pace with social developments and it has become necessary to enact statutes to bring about needed changes; indeed, in recent years statutes have superseded much of common law, notably in the fields of commercial, administrative, and criminal law as well as statutes that reflect changes in technology and science. Typically, however, in statutory interpretation the courts have recourse to the doctrines of common law. Thus increased legislation has limited but has not ended judicial supremacy.

Early common law was somewhat inflexible; it would not adjudicate a case that did not fall precisely under the purview of a particular writ and had an unwieldy set of procedural rules. Except for a few types of lawsuits in which the object was to recover real or personal property, the only remedy provided was monetary damages; the body of legal principles of equity eventually emerged to overcome these deficiencies. Until comparatively recent times there was a sharp division between common law and equitable jurisprudence. In 1848 the state of New York enacted a code of civil procedure (drafted by David Dudley Field) that merged law and equity into one jurisdiction. Thenceforth, actions at law and suits in equity were administered within the same court and under the same procedure.

While interesting in and of itself, what comes from this is the expectation that the government and the legal system to which we are all bound is expected to act fairly and equitably and provide for the codification of behavior. I do not want to diminish or downplay the threat posed by government intrusions given the power that a government possesses to invade a people's property and privacy and, when it deems necessary, withhold their liberty. The so-called War on Drugs, not to mention the War on Terrorism has only made the situation worse and has resulted in myriad abuses in the name of national security. Though not often given to conspiracy theories, the renewed emphasis on national security in the wake of 9/11 has permanently altered the landscape of civil liberties and privacy has been one of the casualties second only to perhaps truth. I will revisit this matter in much more detail in Chapter Six.

Despite the more recent focus on government violations of privacy, times have changed and there has been a paradigm shift in the past 30 years that has given way to focus on the private sector. Governments will always have the power to invade a person's privacy and that threat has not been ignored but the private sector has a very different motive than the public sector; where governments seek information in the belief they entitled to it and require in the interests of governance, the private sector is motivated entirely by profit. Put simply, Orwell's vision of 1984 has been, at least partially, superseded by data warehouses and direct marketing.

The shift in awareness has led to a shift in attitudes and the consumer of the 21st century is a different animal than that found in the 20th century—it is a fundamentally more cynical, distrusting consumer that is more likely to question and equally willing to walk away from the bargaining table. Thusly, history bears witness to an *evolved* consumer—if indeed the economic landscape has become "unsentimental" and tougher then it would appear so to has the population—tactics and techniques that worked well during the 1960's, now fall upon the collective (and selective) deafness of a jaded public.

Concomitant with this battle-hardened populace comes a growing militancy when it comes to privacy and the genesis of the Do-Not-Call-Registries that have emerged in the United States; these lists, which demand companies do not call them, have already resulted in hundreds of thousands of layoffs in the telemarketing sector and this is likely only the beginning. Naturally, there would have been no need for registries if the business community understood that there are social boundaries of behavior and crossing those boundaries has consequences. There is no evidence of a single incident that has caused this backlash but rather the *incre-*

mental increase over time to the point where the sheer volume of intrusions to the average household had become, and remains to some extent, a serious invasion of privacy.

And it is for these reasons, owing to the previous definitions of law, that privacy laws become necessary to set the social and legal boundaries to prevent, and where necessary, provide remedy to those that have been abused. Sadly, every country on every continent has developed a messy mixture of laws and policies governing privacy and the lack of a universal standard can make it difficult to determine what is actually protected under the law. Actually there are universal covenants on privacy (within the larger framework of human rights) and that is perhaps the best place to begin the description of privacy protections and which will set the baseline for evaluating the privacy laws in different countries.

International Standards

United Nations 1948 Declaration Of Human Rights

The easiest point of entry into the abyss is through the international standards and covenants that were drafted as a universal standard that would be basis for the protection of human rights in general. Turning back to 1948, the United Nations signaled a new era in human rights with the signing of the *Declaration of Human Rights* that sets forth the fundamental liberties and protections that were deemed critical to freedom and human dignity. Mind you, in reading the preamble, it becomes apparent that declaration defines more of a standard of achievement that countries should strive for rather than any enforceable act supported by sanctions or remedies. Though comprised of 30 distinct Articles that govern everything from prohibitions on slavery to the enumeration of basic rights such as freedom of speech, the focus of this historical document lies in *Article 12...*

"No one shall be subjected to arbitrary interference with his privacy, family, home or correspondence, nor to attacks upon his honor and reputation. Everyone has the right to the protection of the law against such interference or attacks."

At first glance Article 12 may seem rather dry and it has been argued that it does not fully define the meaning and scope of privacy and it is subject to exemptions but it is quite robust; the Article makes it *abundantly* clear that people shall be free from interference in their privacy, their, family, and their home. The Article raises really thorny questions when it comes to social issues such as child rearing or abortion because it does not explicitly grant exemption on the basis of morality or subjective definitions of the common good; for example the interference in the matter of family would preclude laws that interfere in the organizational structure or dynamics of the family. The inclusion of correspondence within the framework of Article 12 further closes the door to a person's writings and creations that would, if followed verbatim, call into question a majority of police seizures.

Ironically Article 12 serves to provide additional protections against slander and defamation of character that falls under the defense of a person's reputation and honor. Hence, any instance in which a person is placed in a false light as would be the case if an employee's cause of dismissal was made public is a direct violation of the article.

International Covenant on Civil and Political Rights

There are other legal instruments such as the International Covenant on Civil and Political Rights that defines the right to privacy under *Article 17* as...

- *No one shall be subjected to arbitrary or unlawful interference with his privacy, family, home or correspondence, nor to unlawful attacks on his honor or reputation.*

- *Everyone has the right to protection of the law against such interference or attacks'.*

Article 17 of this convention is virtually identical to Article 12 of the former convention. The second clause of this actually refers to fundamental premise of the Rule of Law that states, as noted, that everyone is treated equally under the law and is not directly associated with privacy law. It is a generally accepted principle more so than a specific protection of privacy.

OECD Guidelines on the Protection of Privacy & Transborder Flows of Personal Data

Lastly, drafted in Paris in 1981, though not specifically a law or convention in the strictest definition of the term, the *Organisation for Economic Cooperation and Development's Guidelines on the Protection of Privacy and Transborder Flows of Personal Data* is a more "up-to-date" privacy manifesto that takes into consideration the emergence of computing and the flow of personal information and it is an important standard if for no other reason than it is the foundation for the privacy laws of many countries including Canada. The OECD Guidelines are of substance because they further fuse the acknowledgement of privacy as a human right and the growth of computing—thus extending human rights into the digital realm. The guidelines are about as titillating to read as a dictionary but they do set forth a more contemporary standard of privacy protection and serve as the foundation to many of the directives in Canada and Europe.

Constitutional & National Laws

Just about every country in the developed world, even many that are not at that point in their history, have enacted varying degrees of constitutional and legal protections of privacy and civil liberties that subscribe to the spirit if not the letter of the principles of the international standards. Russia provides protection under Article 23 of the Constitution and I have found myself quite impressed by scope and strength of their legislation. But to stress the importance of a comprehensive constitutional framework for basic human rights and privacy it is easier to look to those countries that possess little or no protections. There are still quite a few countries that have no safeguards to speak of and a look at Singapore is a perfect case in point.

> *"I am often accused of interfering in the private lives of citizens. Yet, if I did not, had I not done that, we wouldn't be here today. And I say without the slightest remorse, that we wouldn't be here, we would not have made economic progress, if we had not intervened on very personal matters—who your neighbor is, how you live, the noise you make, how you spit, or what language you use. We decide what is right, never mind what the people think. That's another problem"*

> **—Lee Kwan Yew, Prime Minister of Singapore, 1986**

Though just a suggestion, I would recommend reading that *again* to fully appreciate why I bothered to write this book and as a reminder that there are many that do not share in an enlightened view of human rights. North American's can, with typical western arrogance, look askance at the ACLU, but the words of Yew serve as a vivid reminder that there really are people, people in power, that do not see freedom through the same eyes. The question to ask then is how much do we differ in North America?

In looking to the United States or Canada, though more specifically the former, there is a temptation to assume that the mantle of freedoms enshrined in our Constitutions protect the population from violations to their privacy or freedom—and while this holds true to some extent, don't let the façade of our culture deceive you—for despite the apparent freedom we enjoy, there are a lot of exemptions to the United States Constitution and the U.S. lacks a comprehensive federal framework that protects privacy—there are simply no *explicit* amendments under the Bill of Rights or the Constitution that guarantee it.

As far as the United States is concerned, the perceived bastion of freedom, even then the U.S. Supreme Court has interpreted a *limited* constitutional right of privacy *derived from* several provisions under the Bill of Rights. Such privacy rights include the right to privacy from government surveillance into any area in which a person has a "reasonable expectation" of privacy. This is a distinction worth noting—this expectation of privacy does not extend to public places or airports (under the Border Exception Rule of the Fourth Amendment) but it would be reasonable to assume that privacy would be found, for example, in a washroom (Katz v. U.S., 386 U.S. 954) but that isn't necessarily so. You are also entitled to privacy against trespasses against your property and matters concerning education, contraception, family relationships, and those domains consisting of intimate facts. And yet there is relentless pressure in the more conservative regions of the country to infringe upon these rights and possibly a move towards their reversal.

There is, however, no generalized privacy law in the United States with the exception of the 1974 Privacy act that provided limited coverage regarding records held by the government. There is also no single agency or oversight department responsible for privacy, which leaves the protection of privacy under a patchwork of legislation at the federal, state, and local level. To complicate matters, every department and agency at every level of government have either developed their own standards or have some direct or indirect influence over the policies that provide some level of protection. And naturally each is protective of its own jurisdiction with the result being a confusing patch work of laws, policies, and exceptions to both.

In three specific instances, the Federal Trade Commission has oversight and enforcement powers over consumer credit information and fair business practices and the FTC, in all fairness, is one of the best government providers of privacy information to the public and they do investigate thousands of cases of identity theft every year. There are other laws that govern the protection of financial records, health records, and the surveillance of all forms of communications under the Omnibus Safe Streets and Crime Control Act of 1967 and the Electronic Communications Privacy Act of 1986.

Ironically, if there is any defense against violations in the private sector in particular for Americans, it lies in of Tort Law and civil lawsuits; even if an organization has not specifically violated any of the many privacy laws, they may be subject to litigation on the grounds of negligence or other act of trespass whether the act was intentional or accidental. The common law is the legal tradition that the United States inherited from England in colonial times. During the last century, American common law developed a body of privacy-protecting theories that give people whose privacy has been invaded the right to sue and collect damages. Notwithstanding attempts at tort reform that have sought to reduce or eliminate both the extent of awarded damages and the right to sue, litigation through tort law is still a preferable alternative to a highly regulatory the environment alone the abysmal failures with self-regulation. The United States, in my opinion, requires a multi-tiered framework beginning with a constitutional amendment and the inclusion of both a federal regulatory framework and retention of the right to sue when these avenues fail to provide sufficient protections.

Despite the proximity of the American cultural influence, privacy legislation in Canada has taken a very different approach that is more in keeping with European standards even though it has taken many years, decades actually, to bring in a legal framework that works across the country at the federal level in addition to provincial and municipal laws that cover privacy. Under the Canadian system of laws, the population is protected under two different legislative acts, the Privacy Act of 1983 and the more recent *Personal Information Protection and Electronic Documents Act of 2001.*

The Privacy Act of 1983 was remarkably limited in its scope and only provided protection against undue privacy violations in the public *sector; the law forced most of the departments and agencies of the federal government by limiting the extent to which information was collected, stored, and used, and disclosed to other parties. The law also gave individuals access to information that was relevant to the person but remained ineffective with respect to the private sector; it would take 20 years before private sector legislation was finally passed. The Personal Information Protection and Electronic Documents Act, introduced over the course of several years, has redefined privacy law in Canada by including the private sector and a set of clearly defined expectations that govern the acquisition and release of personal information. And pivotal to the act was the requirement that individuals have the right to information kept on them and the requirement for consent before information is disclosed. By 2004, the Act strengthened protections by governing the transfer of information across provincial boundaries and included employees that were previously left out.* This was a remarkable change in privacy law in Canada but it is still too early to make any premature judgments regarding its effectiveness—there are many companies that do not even know the law exists and many law firms are just beginning to grasp the nature and extent of the law as it is gradually disseminated to the private sector.

To confuse matters even more so, there are also provincial and municipal laws in every province and major city in Canada, which make it difficult for companies to comply with so many conflicting pieces of legislation. For example in Ontario, residents are protected under the Freedom of Information as well the Protection of Privacy Act and the Municipal Freedom of Information and Protection of Privacy Act. Only Quebec was deemed to have privacy protections of similar strength and as a result, they are exempt from the federal act. I have discussed the federal act in Chapter Seven in a little more detail for those of you that need to understand your obligations under the act(s).

I will close off on my discussion of privacy law by paraphrasing, rather liberally, on the positions of the ACLU that argue quite effectively that the explosive proliferation of surveillance technologies coupled with the erosion of axiological liberties brings closer the threat of a surveillance state. It is worth bearing in mind that North America is comparatively "young", as is our vision of democracy, and has yet to develop the wisdom of hindsight that is accorded to other countries such as China and Russia whose history is measured not in the hundreds but thousands of years. Canada has, with the passage of stringent privacy legislation, proven itself serious on the subject of privacy but the United States has yet to embrace privacy through extensive legal and constitutional protections.

On the surface of it, the law in general and privacy law in particular is probably not on your top list of fun things to read and it can certainly be a frustrating journey but the alternative—a world of no laws or protections—is not an appealing thought. If it was not for the protections we have, whether those laws relate privacy or the underlying framework of protections for all civil liberties, there would be nothing to protect the population against the abuses of ideologues like Lee Kwan Yew—and it would be silly to think that we do not have similar people in the United States and other "democratic" nations. Then again if it weren't for the Constitution, there would be nothing to protect your religion or what you say or what you read.

2.6 The Impact of Cultural Darwinism on Privacy

"The universe is change; our life is what our thoughts make it."

—Marcus Aurelius Antoninus, *Meditations*
Roman Emperor, A.D. 161–180 (121 AD–180 AD)

There have been several occasions thus far when I have stated, rather elusively, that there exists some correlation between evolution and privacy and it is time to explore this idea as a fitting closure to this chapter. Unfortunately the debate on the validity of Darwinism has not abated since the 19th century and if anything it is has gotten considerably worse given the polarization of beliefs between those that adhere to and oppose the theory. This polarization of positions can be attributed to many factors but chief among them is the lack of education in science coupled with a hard ideological shift to the right amongst social conservatives. To compound the problem a majority of the public have never read Darwin let alone Spencer or Wit; and this, paired with the misconceptions associated with evolution, makes my task all the more challenging.

Even as the instigator of this concept, I do not think the correlation between privacy and evolution merits the definition of a hypothesis let alone a theory—instead it is put forth as a gentle suggestion and the origins of a *potential* theoretical framework. But how does one make the quantum leap between the mechanisms of genetic recombination and a social phenomenon like privacy which already defies consensus—it is admittedly quite a conceptual jump but there are merits to the argument. Before we wade into the abyss, however, this demands a brief explanation of evolution, Darwinism, and how biological precepts can offer explanations for human behavior.

Theoretical evolutionary biology *is* a complex field and it is made all the more difficult when mistakes and misinterpretations, even by biologists, find their way into the public sphere. For instance many biologists continue to adhere to the belief that there is a hierarchy of species or a chain of being when Darwin overturned this idea because it was a violation of the principle of common descent. Biological evolution as it was originally defined is really little more than *a change in the gene pool of a population over time* where the gene is the hereditary unit passed on between generations and the gene pool is the set of all genes of a population.

That is a gross oversimplification of biology but, given the scope of the book, it will have to do and a more practical case would provide a better description of evolution than a long-winded treatise. Prior to 1848, the English moth, *Biston Betularia*, was dominated by a light colored variation—so much so that approximately 95% of this species were light. As the Industrial Evolution materialized and the grime from the factories covered the trees, birds were able to spot (and consume) the lighter variation; consequently, by 1898 98% of this particular species of moth were *dark*. The increased proportion of dark moths over successive generations influenced the genesis of offspring (dark:dark) and the resulting gene pool of this population. Thus through genetic recombination and mutation the light gene was marginalized over time.

It is important to understand that it is the *population* that evolves and not the individual. The *individual* moth does *not* evolve from light to dark within its lifetime. And neither do humans—an individual can *adapt* to his or her environment to improve the probability of survival through lifestyle improvements, but these phenotypic changes induced by the environment are *not* characteristic of evolution. Furthermore, we, for that matter any species, are not passive inhabitants—every species modifies the environment as they adapt and every modification results in gradual changes to the ecosystem. Suffice to say human adaptations of the environment including the destruction of rain forests and the increase in greenhouse gases are a permanent reminder of the fact we modify to adapt and we adapt to survive. In the case of the aforesaid moth, there is a causal relationship or interdependency between how humans modify their environment and the evolution of other species—if factories did not emerge, the light colored moth would have survived because their environment would not have been altered.

This is *biological* evolution in its most pure form and it is pure because it possesses the three fundamental mechanisms that define evolution: variation, inheritance, and selection. Accordingly, in the preceding example, the evolution of the moth conformed to these principles. Variation could not have occurred if there were only a light colored species—they would have all perished and their survival required the existence of a dark variant. The moths, over the course of generations, exhibited inheritance as the offspring of parent moths changed in frequency—since they were fewer light moths to mate with the dark colored moth, the resulting frequency of light moths changed over time. And finally the mechanism of natural selection eliminated the lighter moths because they were unable to adapt to their changing environment.

But the English language interchangeably utilizes the term evolution with many words and other synonyms in the dictionary include grow, change, develop, and progress; yet none of these are forms of evolution—they do define some arbitrary change over time but they do not possess the requisite attributes that define evolution. In other words to use the term evolution in any other context than for which it was originally defined is a misapplication of the term.

In fact, academics from many fields of science have made the effort to apply the principles of biological evolution to everything from culture to the development of software and they have made use of Universal Darwinism to achieve that end. Universal Darwinism is an abstract framework that was derived from biology to explain a diverse range of human and computational phenomenon; it is a very contentious theory but I wouldn't be addressing it here if I did not think it was relevant. The fact is Darwin does not mention the gene in his work—Darwinism was developed over the course of 30 years as a *general* framework that could be applied to any complex system. Biology is one *special* instance of the general theory but other special instances can include economics or various elements of the human condition.

There are of course significant differences between biological organisms and any cultural phenomenon and it would be a mistake to attempt to draw a direct correlation between the two systems; it is more appropriate to apply a series of *derived* principles—a way of saying a system is indirectly analogous to a biological system because it possesses behaviors that are similar to the behavior of biological organisms. In order for any system to exhibit evolution, the system must adhere to five key tenets of evolution including a steady rate of change, a common ancestor, variation (or diversity), selection, and inheritance. In attempting to apply Darwinism to economics or sociology it is quite clear that these systems do not display *all* of the required behaviors to be classified as purely evolutionary systems but by abstracting the principles of Darwinism it becomes possible to find metaphorical similarities.

In an effort to avoid the nuances of the debate, I prefer to say non-biological systems do not exhibit evolutionary behavior so much as they *mimic* or imitate these Darwinian principles. Even though it is difficult to superimpose Darwinian principles onto a socio-cultural phenomenon to explain its behavior, there are broad parallels that are observed between the two systems that warrant a metaphorical approach to understanding human behavior. And through the application of concepts such as the Continuity Hypothesis, the cognitive framework for cultural evolution through the transmission of learning and institutions is permitted and which allows a "loose" application of evolutionary precepts to explain human behaviors.

Cultural evolution does have attributes that closely imitate the principles of inheritance, selection, and variation that are required for all evolving species even if the definition of these principles in a cultural context does not include a reference to genetic recombination of genes. In fact cultural evolution possesses several advantages over biological evolution that allow the former to progress at a more accelerated rate than the latter—in other words culture is developing faster than nature.

In biology, there is an inherent randomness to the process of inheritance and natural selection that is not present in human behavior. For example, a somatic protein does not share information with the nucleic acid of DNA but information sharing and transmission is one of the most powerful forces of the human species; when these modes of knowledge transmission are coupled with intentionality and varying feedback mechanisms, the human species are capable, unlike most other organisms, of altering the outcome of (some) events. Cultural evolution achieves the requirement of inheritance through learning and this concept of knowledge transmission within and between generations is central to human progress and survival. As new knowledge is acquired through discovery, this pool of knowledge affects social institutions until the next instance of progress—these advances ultimately result in collective behavioral change that gradually permeates a population.

Selection and variation are also observed within cultural phenomenon although the mechanisms of both differ significantly from natural selection and the diversity of a species. Natural selection is supplanted by artificial selection—a selection process that is affected by the accumulation of knowledge, the ability to make choices between outcomes, and a desire for survival. Consider the case of food intake: in human cultures, the discovery of carcinogenic elements in some foods has increased our knowledge and the assessment of threats to survival that have resulted in changed attitudes towards certain foods that have become institutions. The same can be said of smoking, drinking, and other risks that have altered our perceptions and social norms. Unlike biological organisms, knowledge is shared between generations through institutions to increase the probability of survival. Smoking, once common in offices, is now outlawed virtually everywhere because the knowledge gained through successive generations has been passed on and affected attitudes towards it. This is a *cultural metaphor* of inheritance.

Selection is just as observable in human cultures though it is more appropriate to define it as a form of artificial selection than natural selection. More importantly, humans can apply myriad criteria and thresholds to the selection process that optimize the outcome of decisions. To use sports or the military as an example, though you could certainly use American Idol if you wish, are perfect illustrations of artificial selection. In the latter instance of American Idol, a selection process occurs across the United States to select successively smaller pools of candidates from the total pool of candidates. During this process, a series of "metrics" are applied against each candidate—and like evolutionary processes, either the candidate succeeds in meeting these criteria such as pitch, tone, and the subjective definition of talent, or they are eliminated. American Idol, I should stress, is not an evolutionary process—it does possess variation (different types of singers, and it does have a process of elimination. But it does not exhibit the requisite characteristics of inheritance—the talent of successive seasons is not passed between so-called generations.

Economics, as an open complex system, also possesses seemingly evolutionary behavior and there is an entire field devoted to Evolutionary Economics—and for what it is worth, even they cannot agree on whether Darwinism is applicable to economic systems. Without delving into considerable detail, economic systems do appear to imitate some elements of Darwinism and there is evidence that these systems imitate the mechanisms of selection, characteristic diversity, and inheritance through knowledge transfer. In economic evolution, firms are loosely described as "species" and these species compete for finite resources such as labor, raw materials, and most importantly, consumers. Where it differs is the sheer volatility and speed of transactions—where biology requires a stable epoch, economic systems are characterized by rapid time scales.

The most important component of economic evolution, at least within the scope of this book, however, is the competition for scarce resource and the manner in which economics is reshaping the planet. The development of condominiums in a natural wildlife preserve is a single but stunning metaphor for how human economic behavior is altering the ecosystem; remember that a species does not exist as a passive participant in an environment—every species modifies or has a direct impact on its environment. Similarly, two species cannot inhabit the same environment indefinitely; this is a theoretical concept that has been observed time and time again. Excessive fishing coupled with human population growth has displaced some species of fish and others are extinct as a result of our impact on the environment. The decision by the United States to not participate in the Kyoto Accord to reduce green house gasses is completely predicated on economics and not science.

The most profound force of evolution is the driving desire for survival, which Darwin called the struggle for existence, and this struggle is clearly observable in economics. In biology, it is accepted that the total number of species cannot exceed the total amount of resources; this constraint does not apply in economics and it is quite common to find a large body of firms (species) competing for the same population of consumers—and it is this disparity that results in the struggle to exist amongst firms and the quest for optimization and human creativity. On the darker side of humanity, though, this same struggle has allowed firms (any organization) to adopt predatory practices to sustain their existence. Obviously a firm that has a higher cost of labor, inferior products, or uncompetitive pricing strategies will not survive against a firm that has optimized one or more of these attributes. This process of optimization, however, when combined with the mentality of Survival of the fittest, is the very thing that has resulted in sweatshops and the exploitation of child labor. Likewise, the more recent rise in "global outsourcing" is a veiled expression for cheap labor. Why pay $22.40 an hour in Wyoming when you can pay $0.80 an hour in China. The firm employing Americans would be hard pressed to compete against the former and cost of resources will inevitably lead to their extinction. The same attitudes towards cheap labor is crossing over in the *social* space where the quest for the "perfect employee" and the elimination of "undesirables is having an dramatic impact on our culture.

That is where many of the more *dangerous* (and intentional) misapplications of Darwinism come into play.

As the economy has changed (or evolved if you prefer), and I must stress they are not always positive changes, increasing competition and the erosion of social values is giving rise to a different society than we had in the 1950's let alone the 1850's. As economic systems continue to move faster and organizations become larger, the ability for leaders to impart and sustain a coherent collective vision becomes more of a challenge result in reactions that include the adoption of both passive and active forms of supervision and surveillance as a feedback mechanism. Suffice this is not observed in all firms and the extent to which a firm engages in surveillance is a function of the corporate culture. Equally desired is a collective predisposition that favors a common organizational goal rather than private self-interests. I question the validity of any economic theory that posits an individual will abandon their interests to embrace a collective goal if that goal does not advance the position of the individual. Every organism will seek to further its own interests and will only engage in cooperative behavior when it is in the interests of the organism. As economic systems continue to exert more influence on the social sphere, such influences have resulted in the pursuit of Social Darwinism. Let me start by saying that, even though it is not entirely invalid, Social Darwinism is a risky enterprise because it has tended to misapply evolution to justify highly questionable ethical behavior and decisions—in other words, some within this field have misapplied science to rationalize their belief systems. Social Darwinism, based on the work of 19th Century philosopher Herbert Spencer, is a rather brutish interpretation of Darwin's Fifth Principle, natural selection.

First and foremost, the intentional misinterpretation of biological evolution as a social doctrine has provided a framework for discriminatory beliefs and philosophies that included the assumption that the European male evolved faster than other races, the genesis of Nordic Racism, and eventually Nazism and Eugenics. It is fair to put forth that the Holocaust was largely the product of nazi theoreticians that reshaped the tenets of biological evolution to justify the extermination of a species and therein is the inherent danger of hijacking science to suit a specific philosophy.

Just as nature eliminates weaker organisms, the adherents of Herbert Spencer's theories held that the same should apply to the human race and that the unfit should "cleansed" away. When applied, erroneously, to economic theory, there was a belief that concepts such as welfare or the feeding and housing of the poor permitted the survival of a weaker species that in turn weakened the human gene pool. Contrarily, it has been theorized that the fittest of the species are those with more money and power and this in turn led to the belief of Spencer that the concentration of wealth was a positive force. And far from an anachronism of the Industrial Revolution of the 19th century, these philosophies continue to persist today. The application of Social Darwinism is not just a theory but a spurious ideology that has permeated American culture in the form of an eroded social safety net, the weakening of civil liberties, and myriad instances of cultural decay; even class warfare and the gradual demise of the middle class are glaring examples of this tectonic shift in ideologies.

Though I do not want to detract from my discussion of genetic monitoring discussed in chapter four, it is quite alarming to think that a presumably autonomous individual must submit to invasions of the human body to satisfy the any arbitrary dogma and allowing such "isolated incidents" will open a Pandora's box within ten years. For what begins as random drug testing of police officers or pilots has the potential to become mandatory drug testing of an entire population which will eventually give way to the genesis of Socio-Economic eugenics—and there is ample evidence to support the potential shift towards a neo-eugenic movement. In several organizations, smokers were terminated for refusing to either quit or give in to drug testing. Whatever your position on smoking, what prevents the same doctrine from being applied to those with diabetes, cancer or any condition deemed a drain upon society. Insurance companies already eliminate those with pre-existing conditions now and there is little to prevent organizations from applying the same ideology.

Alas I return to the subject of privacy and of greater import its relevance to evolutionary theory, particularly economic Darwinism. It would stand to reason that those that question the value of feeding the poor or providing housing and adhere to the concentration of wealth will be the same to question the validity of privacy and freedom. After all if, as they believe, that the concentration of wealth is for the common good, then surely the elimination of the poor and weak are also for the common good. I am inclined to suggest that the term "common good" has is a veiled reference to any circumstance in which the few profit at the expense of the many. The absence of civil liberties and a universal code of human rights would undeniably make life far more miserable as would the elimination of constitutional rights. The framers of the United States Constitution understood that individuals should have the right to practice their own religion or express themselves even when such beliefs ran counter to those of government. It is also implicit that these protections guard against dogmatic abuses of the government and we are seeing this occur more frequently with respect to reproductive rights.

Insomuch as privacy is concerned, the suggestion that I make here is that privacy exhibits many of the same characteristics of Darwinism and more importantly attitudes towards privacy are directly influenced by cultural evolution. The gradual erosion of privacy has diminished the individual's ability to prevent instances of artificial selection in which personal information is used to make decisions that are anything but benevolent. Privacy, and by extension self-determination, is quite simply a constraint on the doctrine of Survival of the Fittest. Privacy, then is, just one of many defense mechanisms that guard against decisions that affect our right to not only be left alone but survive. If for example, an organization was to impose mandatory drug testing that may reveal a genetic anomaly, and you were to possess such an anomaly, then such testing constitutes a *threat* to your financial survival. It's a defensive shield that works in concert with other civil protections, to guard against harm.

Hence the declaration of human rights, including privacy, regardless of the source of such declarations is mankind's way of counterbalancing the harshness of man's darkest abilities and inhibitions. In the absence of child slavery laws, North America would be no better than any other third world country—and it is these protections that differentiate a democracy from a totalitarian regime. But lest anyone enjoy a brief moment of arrogance, democracy does not necessarily beget benevolence and there are some that would argue that transnational corporations have taken the helm from government and hijacked the social agenda.

The application of evolution to the social sciences, despite its deficiencies, remains the best explanations for the human condition and by extension it offers at least a partial explanation for invasions of privacy. Even if the theory of evolution was invalidated as a universal abstraction, it is still the originating *cause* of human behavior. It is, though, only a partial explanation and does not account for the environmental factors that influence privacy including technology. Nevertheless, it is, so far, one of the most compelling explanations of invasions of privacy.

2.7 Undercurrents of Humanity

"True happiness is of a retired nature, and an enemy to pomp and noise; it arises, in the first place, from the enjoyment of one's self, and in the next from the friendship and conversation of a few select companions."

—**Joseph Addison (1672–1719),** ***The Spectator,*** **March 17, 1911**

As our culture has become increasingly obsessed with materialism and wealth in addition to an equally blind devotion with technology, there is a habit of equating human existence in either economic or technological terms without giving due consideration to the psychological and sociological influences that guide human behavior, especially as it relates to privacy and respect. That is not to condemn the accumulation of wealth or the advancement of science and I am just as given to the acquisition of wealth and a healthy admiration of science. But it is the underlying human condition and aggregate cultural norms that dictate attitudes towards privacy and they are often neglected in discussions of privacy.

The impact of science and technology on our culture is crucial to our progress and survival but it is only *one* dimension of existence. And is this failure to recognize the social dynamics that is partially to blame for the erosion of privacy? There is more of focus on so-called Privacy Enhancing Technologies than the underlying causes that require these innovations to exist in the first place. Rather than blaming it on code, the onus rests upon the lack of a collective conscience. But it is not entirely fair or accurate to place the blame exclusively on the entrepreneur or engineer when the general public is just as responsible for their apathy and governments held to account for their cynicism. Attitudes towards privacy and respect for the individual in general, as far as I am concerned, are seeded during the formative years of childhood and those attitudes continue to be shaped by an individual's environment as they mature through adolescence and adulthood. Ergo, childhood experiences that give rise to selfish behaviors will manifest themselves in adulthood.

These psychological factors not only influence personal growth and self-perception but they have a marked impact on social intercourse—after all, if there was no need to form bonds with others, privacy would never be an issue, but as social animals interaction with people is central to our culture and for that reason boundaries become necessary. Whether involved in a person-to-person interaction or within a group dynamic, the relationships or more importantly the depth of a relationship defines the extent to which we are willing to trust a person and with that the boundaries of privacy are changed as well.

How a person picks a friend or partner is influenced heavily by a series of criteria that define the *type* of people that we would like to be with and these attributes are shaped by geography, social status, upbringing, education, and income in addition to many other factors such as sexuality and character. But it is equally vital to mention that the criteria for selecting a mate or friend can change over time. For example, if you have not heard from someone in a long time that was once considered close then that is the effect. The cause, barring any incident that may have resulted in a strained relationship, could be a change in the person's priorities or what is important to them. In other words you might have fallen off their "radar" if their life has changed in some way. I rarely accept the argument that a person is "too busy"—that is a merely a euphemism for losing interest and if they are still interested then they will find the time—friendships of any kind require work and an investment of time. Marriages also undergo changes and what might have been wonderful in the beginning of the courtship may seem trivial or annoying five years hence. That brings me to some closing thoughts on trust because it is related to the depth of a relationship.

If a relationship is not predicated on trust then it is just an acquaintance at best and will never reach a meaningful level in the absence of trust. If two people do not invest the time and energy to discover each other than it is unlikely the friendship will grow let alone survive the test of time. Likewise, I have argued earlier than trust is loosely symmetrical and reciprocal that ties to my thoughts on trust equilibrium—it is always a two way street. Contrarily, the trust between individuals and organizations is almost always asymmetric and based on a calculus of risks and expectations. We trust organizations to do what they are paid to do and nothing more. Trust is a gateway beyond the barriers of privacy that are erected to prevent disclosures to those we don't trust and for that reason trust must be respected and betrayals are rarely if ever repaired.

With a bit of luck and argumentative arm-twisting, hopefully your attitude towards curiosity has changed a bit. Curiosity is a gift of nature and is possibly one of the least appreciated abilities—Einstein was, contrary to what you may think, not a mathematician, it was his creativity, imagination, and curiosity that gave birth to relativity as well as his lesser known works. If it wasn't for our curiosity, there would be no fire, no penicillin, and no space shuttles—we'd still be banging rocks together. Curiosity plays an important role in relationships too and the evidence is overwhelming—couples and friends have a heightened sense of curiosity and the need for discovery also have stronger bonds. The tricky part is determining when curiosity is benevolent or harmful and determining the boundaries of discovery.

It is the interplay of curiosity and trust that determine if or when private facts become public domain and the degree of trust you have in a relationship is based on your belief that the person will not betray that trust and disclose things that were meant to be kept confidential. Gossip is a significant component of cultural development and communications but my focus is on the disclosure of personal information and in that context gossip is seen as a negative force. Gossip may have originally served as a means of gathering information about food and shelter, but this is the 21st century and the nature of gossip has changed over time—not so much in how it used but what is said.

In light of the tendency of people to gossip coupled with the understanding that gossip is hardwired into the human psyche, it demands that the same under standing of secrecy gain the respect that is rightfully deserves. Gossip and secrecy are two diametrically opposed forces than remain forever in tension; just as we will never be able to stop gossiping, it is equally reasonable to expect that individuals and groups would not want to disclose facts that might prove damaging or embarrassing. Given this addiction to gossip and the corollary misconception that the public has a right to know everything, it is no wonder that celebrities and public figures take substantial measures to protect their privacy. If it were not for photographers sticking their lens through the backyard gates, there would be no need for electric fences. Buzz!

As for organizations and economics, it is well known that people act differently when in a group than they do as an individual and that, coupled with the forces of economics, leads to behaviors that are not always conducive to privacy. Just the sheer pressures of competition alone force organizations to adopt practices and policies that run counter to the very same principles that we expect as private citizens—it is a type of social paradox which can be explained by our instinct for survival. It may not be attractive, but it is logical.

These economic forces and the resultant organizational behaviors are the very reason that a strong legal framework is required to prevent predatory businesses practices. There are laws to punish those that prey upon children and to protect future generations from this behavior; there are laws against insider trading because insider trading demonstrates a lack of integrity and fairness—all of these "laws" are defensive mechanisms that counter the forces of Economic Darwinism.

And for the reasons that we need laws to protect against human rights abuses are the same reasons that we require strong privacy protections and it is somewhat encouraging to see Canada and Europe take it more seriously as they adopt a more comprehensive statutory framework. Basic concepts such as consent and collection amongst others have finally been incorporated into the laws of Canada and Europe that will place the onus on governments and the private sector to abide by these minimal standards. It is still an unfortunate reality that privacy has not been incorporated as a constitutional right as it ought to have been but progress is being made.

Insomuch, as the relationship to evolution and Darwinism, well that is something quite different. To accept this correlation requires not one but three distinct quantum leaps of thinking and it is my expectation that approximately half of all readers will be able to make that jump. First and foremost is the acceptance of Evolution itself—and if your religious or philosophical upbringing forbids this belief than the second and third become meaningless. Secondly, you must also be able to accept the principles of Universal Darwinism and the abstraction of Darwinism to all open, complex systems. And finally, you must be willing to consider the possibility that privacy is a defense mechanism that guards against predatory behavior. I have no intention of getting into a debate on the viability of evolution as a theoretical framework or to engage in the endless battle between Evolution and Creation—that is best saved for a different forum. Even if you do not countenance the existence of evolutionary change you can still get a lot from the chapter while disagreeing on one segment. A disbelief of one segment does not extend to the others.

Chapter Two, though a marked departure from the rest of the book, is to my way of thinking, one of the most important passages that counter-balances the collective focus on technology. For that reason, I had hoped to capture, even superficially, enough of the social dimensions of privacy to bring home the point that there is so much more to this than computers and DNA. It is important to remember that technology is only a tool, an important tool, but it is still just that. Invasions of privacy do not start with Data Warehouses, they start with your neighbors. And as the journey progresses into the technical chapters, the social causes will provide the infrastructure for understanding why we have become the objects of unwanted institutional attentions and the connection to things like genetic testing or biometrics.

3

Privacy & the Internet

3.1 Monsters and Miracles

"He who fights with monsters might take care lest he thereby become a monster. And if you gaze for long into an abyss, the abyss gazes also into you."

—Friedrich Nietzsche (1844–1900)

Despite personal misgivings regarding the manner in which the Internet is characterized, reported, and glamorized far more than it is entitled to, the Internet has undeniably become one of the most powerful forces of social change since the invention of the printing press or the telephone and it should not be that difficult for readers to respect the impact it has had on society thus far and the unchartered future that lies ahead. Of course this is only the beginning and it remains to be seen how the Internet is going to unfold in the next fifty years. But whichever way this goes, you may be assured it is going to impact you at some point in your life and privacy will be one of the crucial battles that loom ahead in the next few years.

Sadly, and I suspect it is largely the product of our obsession with technology, it is far too easy to be distracted by the darker elements and this omnipresent "gadget fetish" while ignoring many of the creative endeavors that are rarely reported but that far outweigh the negatives. It is also too easy to take the other path and dismiss or condemn the excesses of the Internet based on beliefs that are based on fallacious arguments and quasi science. And somewhere between these two extreme ideologies lies the truth—the Internet is neither a savior to mankind nor is it Dante's Inferno.

For these reasons there must be a more objective means of differentiating and evaluating both the positive and negative elements that are common to most human endeavors and the Internet is certainly no exception. And in so doing it is hoped that speculation and sensationalism are replaced by a more balanced, and scientific approach. Fair enough, this is not the sexiest approach but technology isn't sexy, its function is to solve problems, something that is often forgotten. So let's start with a candid examination of some of the problems.

The spam that inundates your e-mail is prima indicia of *one* of the most serious crises to plague the Internet and which undermines the positive experiences that the Internet offers. Derived from multiple sources including Google, Jupiter, and Harris Interactive, the sheer volume of spam has clearly reached epidemic proportions and it is safe to say it is consuming the Net. This so-called epidemic, according to some of these sources, now accounts for 30-50% of all e-mail and the costs to organizations runs in the billions of dollars every year. It is a perfect example of an abuse of technology that has distorted the public perception to the point where the Net is now synonymous with spam and porn when that is not the case.

Adult content popup advertisements alone act as a sobering reminder that there is a dark, murky underworld that leverages upon this technological marvel and that is the proverbial tip of the iceberg. For what it is worth, pornography remains one of the most successful revenue generators to date and it is still one of the top search request on the major search engines—the Internet didn't create pornography, it just acts as a conduit to satisfy that dimension of the human condition. On occasion I am a little surprised that it receives so much attention given that humans are a sexual species and it is unfortunate when there are more pressing problems like identity theft and fraud. Lest anyone need a reminder, sex sells and that is why it is on the front page and subjects like identity theft do not possess the intrigue or glamour to complete for coverage. But these problems do exist and until they are resolved, the promise of the Internet and e-commerce will never be fully realized.

Conversely, the problems and social ills I've mentioned so far are not limited to the more obvious transgressions and the cultural impact on society may not be immediately visible. It is difficult to get an accurate count of total users on the Net and the estimates vary wildly from the ridiculously low to impossibly high—the Computer Industry Almanac, pegs the Internet population at 945 million people for 2004 and estimates that the number of users is expected to grow to 1.1 billion by 2005 even though this number seems grossly exaggerated. Likewise an increasing volume of voice and fax calls are being routed over the Net so even if you aren't online you are still impacted indirectly by its existence.

It is not only a question of how many people are online or how they are using the Internet, but just as importantly, how is the Internet influencing life *off-line* and for those that have never even used it. Whether or not you ever use the Net is irrelevant—data *about* you still flows across it. Even the use of e-mail and other online communications protocols like instant messaging or Internet Relay Chat (IRC) will gradually alter the nature in which communications occur between people and it further threatens to displace time spent in person. And the more pressures that are placed on individuals, the more probable it is that we'll have to resort to different communications tools just to stay in touch. This is not a blanket assault on technology but please take into a consideration the broader social influences before jumping to quickly to endorse every gadget that comes along.

And since there seems to be an endless tributary of negative news, I would just as soon place an emphasis on the different ways in which the Internet has brought a positive influence to society in general and yourself in particular. Starting off with one of the more obvious effects, the convenience of price comparisons, product research, and just-in-time delivery has altered the meaning of shopping, as it is known—not only in terms of what can be bought but also the relationship between buyer and seller. As for convenience, there is nothing that cannot be purchased at the click of a mouse from tickets to the Bolshoi Theatre in Moscow to a fully functional F-18 Hornet Fighter for $9 Million that was sold on E-Bay. The Internet has also changed how consumers make decisions and this is certainly evidenced by revolutions in the travel industry—decisions that were once based on the recommendations of the travel agent can now be substantiated by traveler reviews, site visits, aerial photography, and satellite imagery.

However, that is barely scratching the surface and only speaks to a few of the things you have likely read about or used yourself; there are countless benefits and applications that are available on the Net and that are usually invisible to the public. As a research tool, the Internet is absolutely indispensable! Google alone offers more than 4,285,199,774 web pages for your mind to sort through. Even assuming you could read a page a minute 24 hours a day, it would still take you more than 7,000 years to get through them all and that is just the pages indexed by Google. While doing my doctoral work the Internet proved one of the most valuable tools in my arsenal, allowing me to find obscure 15th century original Latin passages of the *Steganographia Trithemius*. And now Google is in the process of developing the technologies to index millions of pages of academic papers that will accelerate scientific research by making information more readily available anywhere in the world.

Actually, the Internet was originally meant to be a research platform that would allow scientists to share and collaborate in real time on the very things that, at some point in the future, will change the world. There are myriad examples to draw from in every field of science with my favorite being the Human Genome Project that is sponsored by the U.S. Department of Energy and the

U.S. National Institute of Health that offers the entire map of the human genetic structure online to anyone that wishes to see it. In another instance of medical research, surgeons in one part of Canada can perform a variety of surgeries through tele-robotic surgery such as laparoscopic cholecystectomy, Nissen fundoplication, Cardiac and thoracic surgery for mitral valve repair, atrial septal defect repair, and internal mammary artery (IMA) mobilization for coronary artery bypass grafting (CABG).

According to the Ontario Ministry of Health…

"…Long-distance tele-robotic surgery may offer specific advantages for remote novice surgeons during their early experiences with minimally invasive approaches to surgery. For example, few urologists have substantial experience with laparoscopy, and laparoscopic radical prostatectomy is a very difficult operation to perform."

Mind you, it helps to look for the fine print…

"Potential intra-operative hazards are possible. During surgery the instruments held by the robotic arms may collide inside a patient's chest or abdomen or there may be interference between the robotic arms moving over the patient."

Though not seeking to underestimate this particular advance in medicine, the Ministry of Health would find the public somewhat more receptive if they would refrain from using the words novice and surgery and radical prostatectomy in the same sentence. Nor does the expression "collide inside a patient's chest" inspire public's confidence. Despite the attempt a little bit of dark humor, these are incredible developments in medicine that will save a lot of lives and the underlying infrastructure of the Internet is making it possible. It is far more complex than I am letting on and obviously surgeons are not using Internet Explorer over dial-up—these are highly secured dedicated redundant lines that allow the surgeons to perform procedures in real time—still the Net has been instrumental in its development.

In other cases such as automotive engineering and retail operations, suppliers and manufactures can achieve extraordinary operational efficiencies by laying secure Extranet tunnels over the existing Internet infrastructure and leverage the Internet to reduce costs or collaborate on product development. Even Voice Over IP (VOIP) is gradually beginning to see the light of day and eventually we will be able to use the Internet for both local and long distance phone calls.

And lastly, as a sign of things to come, some cities like Philadelphia have begun to offer inexpensive wireless access through the city at half the price of high speed access provided by the carriers; and despite a desperate bid by the aforementioned firms to prevent this from occurring, the trend will continue and provide affordable access to the masses in any city in North America and perhaps, eventually, the world. As that prospect begins to unfold, it will become even more crucial to gain a better understanding of this strange universe and the impact it will have on all of us; and what better place to begin than by taking a brief look through history.

3.2 A Brief History of the Internet

"Technology is a way of organizing the universe so that man doesn't have to experience it."

—**Max Frisch**

Just as inspirational leaders are few and far between in any field the same can be said of the visionaries that have reshaped the word, not the engineers and technicians but the rare breed of mind that possesses the unique blend of computational and creative genius. For what it is worth this type of genius is rare even amongst those that are deemed brilliant—Stephen Hawking would be in that class as would Einstein, Darwin, well you get the idea. And the computing field is filled with people like MIT's Marvin Minsky, the father of AI. It is their imagination and creativity that make the impossible possible and the Internet was once considered precisely that.

Needless to say the Internet that everyone takes for granted today began in a very different era and a brief walk through the history of the Internet is worthy of your time and it will set the stage for a discussion on how to enjoy it without becoming the victim of its excesses. Though visionaries had previously considered a global network for communications, it was the Cold War that provided the catalyst that brought that vision to fruition.

At the time, President Eisenhower, responding to the October 4th, 1957 Soviet launch of Sputnik, came to recognize the need for a more appropriate research agency and thus was born the Advanced Research Project Agency. On February 7th, 1958 the Department of Defense issued Directive 5105.15 that established ARPA with the mandate of researching and developing computer and communications technology that would give the United States a decisive, strategic *military* advantage. Do not lose sight of the military focus of the times and that the Internet was not availed to the public until the early 1990's.

Nevertheless, the Internet really got its start during the 1960's when Dr. J. Licklider was selected to lead ARPA's (eventually DARPA) research into the military applications of computing. Considered a visionary, he had established the early foundations of the ARPANET that was a predecessor to the Internet; and for what it is worth, the first demonstrations of these earlier ancestors were quite humble by contemporary standards of global computing. That first generation consisted of four "nodes" connected over a comparatively slow 50 Kilobit connection between SRI, UCSB, University of Utah, and UCLA. Without getting into the heavier details, it still took the invention of packet switched networking at M.I.T. and the eventual adoption of TCP/IP before the Internet was truly born. But these historical developments provided the initial infrastructure that would eventually grow to become the "Net".

Through the early 1970's, many other large academic institutions came online including MIT and Harvard in 1970, Stanford and Carnegie-Mellon in 1971, and NASA soon thereafter. I've left out many of the other "big" players during that historical period so I apologize to purists everywhere for my brevity and for taking literary liberties with history—it is to establish the fact that the network was gradually stabilizing and becoming an important feature across many of the greatest universities and research institutions across the United States (and elsewhere to a lesser extent).

Another by-product of the Cold War was the ability to survive multiple failures and yet still function. During the 1960's and 1970's there was a justifiable concern that a nuclear attack would incapacitate the communications capabilities of U.S. military forces and make it difficult if not impossible to respond. To solve that problem, the infrastructure of the network was developed so that traffic could be re-routed and delivered to their destination. Allowing routers to have equal authority meant that there would be no single "choke point" that could bring down the entire ARPANET (Internet)—military communications could then continue in the eventuality of a *hot war*.

There were many other developments including the adoption of TCP/IP by the Department of Defense in 1980 that replaced the older Network Control Protocol (NCP) that has given the Internet both stability and scalability. Eventually the NSF created NSFNET that formed the national backbone and which also signaled a dramatic shift from military-centric uses to educational use. The military eventually moved to their own networks such as MILNET that is part of the U.S. Defense Data Network, the Non-Classified Internet Protocol Router Network (NIPRNET) and the super secret SIPRNET (Secret Internet Protocol Router Network).

Anyway, a network alone is utterly useless if it does not have software to enhance or make more productive the lives of those using it and it took quite some time for applications to proliferate throughout the Internet to the point where they became universally accepted protocols; what we now take for granted with e-mail actually wasn't created until 1972 and even then the concept of "user-friendly" would be quite comical when compared to what you have now—almost all of the software was predominantly command line and demanded a fluency in unusual command line dialects.

But you must keep in mind that the primary users were not executives but instead consisted of engineers, scientists, and software engineers—the operating systems and languages were developed by and for the technical elite. Other protocols such as FTP came about in 1973 followed by more applications like WAIS and GOPHER servers that were essentially text and menu based search engines for FTP sites and textual documents. No graphics, no sound, and no pop-ups. USENET, though not originally a component of the Internet, eventually allowed for collaboration through threaded discussions in what was referred to as a newsgroup. Though it has grown to unimaginable size, and though the nature of many of the 100,000+ groups is highly questionable, there is still a lot of fascinating social and scientific collaboration work that continues to this day.

Prior to that point in time when the public finally gained access to this global network, one of the most important developments was the invention of the browser. Created by Tim Berners-Lee at the European Laboratory For Particle Physics (CERN), the protocol outlined a new means of information sharing that would revolutionize computing as it was known at the time; the World Wide Web finally took shape in 1991 and was followed by other innovations such as MOSAIC at the National Center For Supercomputing Applications (NCSA) in

1993 and eventually Netscape and Internet Explorer. Many of you reading the book have only been exposed to the later generation of browsers so it might be difficult to understand the gravity of this advancement and the impact it has had on technology and culture. It would be analogous to asking the wealthy to appreciate what it is like to be poor—unless they've been there and lived through that experience it is virtually impossible for them to share in that experience. The same is true of both Internet neophytes and veterans—it's hard for a new Internet user to fathom what it would have been like living in the UNIX or VMS world if their only exposure has been Windows XP. And yet, to offer some balance, I and many like myself miss dearly those early times and still look very fondly upon UNIX. Indeed for many in my field UNIX is still the only operating system of any worth.

It is important for veteran Internet users that have been around for many years, possibly decades, to realize that the average person only came online in the mid to late 1990's. Delphi, one of the early service providers, finally began offering e-mail service in 1992 and was eventually followed by the other giants such as AOL, Prodigy, and CompuServe. Even then, it wasn't really full network access but dreadfully restrictive gateways that provided specific services like mail and news. By 1995 and on, there was a massive proliferation of "true" Internet Service Providers starting to springing up around the world and the number of hosts grew exponentially over the years. Interestingly, the birth of a public Internet ostensibly signaled the end of the BBS era.

There is still the occasional debate on the merits of allowing the public to the access the Internet and presumably there will always be the ardent few that feel it was a terrible mistake—and in the face of spam, spy ware, and all of the other problems, you can appreciate the validity of their arguments. Still, even in spite of these problems, the majority of scientists and veterans of the community believe that technology must benefit society and not remain the exclusive domain of the technocrats and elite. I am inclined to go further and argue that the competitive and social advantages of the Net merit subsidized access for the entire country that would provide global access and eliminate a generation of "Technological Have Not's". I am sure the telecommunications industry will love me for that position.

The rest of Chapter Three focuses on the more pragmatic threats that you may encounter online and countermeasures to assist in eliminating or at least mitigating those risks. It does not necessarily have a direct bearing on privacy but improving your security practices can surely reduce the probability of becoming a victim. And given recent estimates that over ninety percent of all PC's are infected with at least one spyware infection, perchance the inclusion of a chapter on Internet Security and Privacy was not without its merits.

3.3 Understanding Online Threats

"On The Internet, no one knows you're a dog"

—New Yorker Magazine

Regardless of where you are, the unfortunate fact remains that a person is subjected to a broad range of hostile threats whether he or she is in the parking lot or shopping on the web—the medium may be different, the risk of violence may be reduced, but it does not entirely eliminate the range of threats that a person faces; in fact old threats have been re-engineered and new threats have emerged. What has changed has been the wide spread proliferation of the Internet.

While the Internet does not create crime, its very architecture can make it difficult to determine who you are dealing with and what their motivations are—even visiting web sites without the right defenses can prove ruinous. The Internet also provides a vehicle for those bereft of principles to create any illusion they wish to further their own aims. If they seek to damage, they no longer need to break windows, crashing a server or stealing credit card numbers will do just fine. If they seek to defraud, the Internet provides a wonderful place to hide. The important point is that while the technology has changed, people haven't. They have basically adapted to a new world, and if readers are to cope with this new world it will require have to learn the same ropes.

And it is this inherent ability to do as one pleases, to say as one wishes, and to travel as if somehow anonymously, undeniably the very things that make the Internet so liberating, can also be used to the detriment of society; criminals have seized on the Internet to ply their trade with as much success as they have had offline. They were able to learn quickly, perhaps faster than the average user, that the Internet presented new opportunities that would have been impossible using conventional media and tactics, and it is much easier to run and hide than ever before, thus making this technology a criminal paradise.

This applies to any fraudulent or criminal enterprise and includes everything from fraud to child exploitation; in the past year alone several American firms arranging sex tours to Thailand were closed down and their principals arrested after it was discovered the companies were blatantly advertising the tours. The Internet did not create the sex trade but it does give it a global reach.

There are many other forms of threats that have taken a new twist including pedophilia and sexual assault; predators have discovered an untapped wealth of potential targets that were not availed to them in the past. Now they can establish relationships over a longer period of time, utilize social engineering tactics to engender trust, and set more elaborate traps. It is this pseudo-anonymity that has emboldened attackers, regardless of their objectives, to use the Internet as a proverbial hunting ground.

In light of these particular threats there will inevitably be those that demand changes that would rob readers of the ability to remain relatively anonymous but that is not only the wrong approach, it is also playing into the hands of those that seek to control what you do and where you go. While a purely anonymous scheme has its share of problems, partial anonymity is a pivotal function of privacy on the Internet, or for that matter anywhere else. If were not for that simple idea, everyone would be able to see where you went and what you did—though presumptuous, I doubt if that is what anyone would want. The current infrastructure allows for the tracking of a person right to their home or office without additional development.

Whether an individual seeks to steal information, vandalize computers, defraud the elderly, or stalk a celebrity, thieves and others use common tactics to achieve their goals. The predator engages in forms of persuasion either by marshalling the direct route involving logical, systemic arguments to stimulate a favorable response or the more peripheral route that rely on short cuts through the logic using emotional triggers that seek to avoid critical thinking processes. The latter relies more on deceptive practices than the former but both are quite effective—and the Internet makes these practices much easier to accomplish.

The most common form of persuasion seeks to influence a person by appealing to our most primitive emotional states such as greed, ego, ideology, fear, or the need for acceptance. Looking to defraud, a criminal will always attempt to bypass the logical arguments and our natural skepticism by offering large cash prizes or incentives that appeal to our sense of greed and material lust. Business opportunities that offer fast returns are appealing to the need for immediate gratification without effort. Health scams offer cures that soothe our worst fears of death or physical acceptance. Whatever the end goal, the methodology seeks to elicit a very strong emotional response that distracts and interferes with our cognitive capacity for analysis and the ability to counter the pitch. These appeals to our most primitive needs are usually coupled with sophisticated marketing and site development that make it very, very difficult to differentiate between what is real and what isn't.

Whilst only indirectly relative to privacy, the proliferation of Internet scams and scandals very much alive and each scam necessitate personal information for their success and the fact these scams are still ranked in the top ten frauds according to the U.S. Federal Trade Commission warrants their inclusion.

1. Internet Auctions

With the growing popularity of online auctions there will always be an increased risk that you might be defrauded in the course of a transaction. There are many reported incidents of people not receiving merchandise or receiving an item of lesser value than was purchased. The only protection is to exercise due diligence by checking into the sellers reputation and that of the auction itself. Even on E-Bay there are instances of fraud that occur with alarming frequency.

2. Internet Access Services

Essentially this scam offers consumers "free money" to cash a check and switch Internet providers. The catch, though, are very complex long-term contracts that can be difficult to cancel or void without punitive charges for early termination. Watch carefully for any unusual charges on your phone bill and your credit card. Best of all, don't respond.

3. Credit Card Fraud

There are hundreds if not thousands of credit card fraud schemes that have made the transition from bricks and mortar to the Net. In the context of online surfing, most of the scams require a credit card number for "verification" of your age or other information. Once the operators have your card, it is not long before charges will be run up in your name. As a general rule, don't use your credit card except with merchants that you are comfortable with. You may wish to talk to your bank about fraud coverage and their policies governing unauthorized transactions. If this seems as if it ought to be common sense then it would not continually be listed as one of the key security problems—there are a lot of people that do not do their due diligence.

4. International Modem Dialing

I haven't seen this in quite some time but it still being reported with a sufficient frequency to justify a Top 10 placement; not everyone in North America has access to DSL and high-speed connections and modems are by no means obsolete just yet. In this scam, victims download a "free" dialer, which then disconnects from the local ISP and calls back on a long distance line. The same cautions apply to credit card frauds—read your statements and avoid anything that looks too good to be true.

5. Web Cramming

This is an interesting scheme that offers free web sites for a 30-day trial and then charges the consumer additional fees and charges that were not agreed to—even if the consumer never signed a contract.

6. Multi-Level Marketing (MLM) and Pyramids

By no means new, MLM schemes are now able to reach a much larger potential audience than was previously the case and it's ranking as a top complaint indicates there are many that are still falling prey to these scams. In MLM, you are usually promised exceptionally high revenues or returns for items that you sell. What they don't tell you is that you're selling to someone else that is already a member of the MLM scheme. If you come across an offer than involves the purchase of inventory or recruiting others, run.

7. Travel And Vacation

Well of course I'd love a trip to Cuba for $199; but I am somewhat concerned what kind of aircraft is going to get me there if at all. Travel and vacation scams offer ridiculously low prices and then either deliver sub-par accommodations or no vacation at all. If offered a spectacular cruise, it would do well to make sure it isn't the Titanic.

8. Business Opportunities

If you were inclined to consider the trip for $199, there is a wonderful business opportunity just waiting for you. Not a day goes by that someone isn't convinced that they too can be next Donald Trump. I'd like to live like that too but it takes less than a second to realize you won't get a Lear Jet by filling envelopes. If a business opportunity, such as a franchise, is legitimate than they will have the history and reputation to back their claims not to mention other business owners that can validate the claims.

9. Investment Schemes

I need not belabor the point on investment schemes; they are remarkably similar to the business opportunity frauds. In either case, your desire to succeed is being manipulated and sadly has clouded the judgment of many that have lost significant income in day trading schemes and many others. Deal with a licensed broker with a reputation in your community or some of the better known and established online brokerages.

10. Healthcare Scams

I think this is amongst the most unfortunate of scams and I've seen evidence of these schemes in e-mail and other Internet channels. The worst of this is the hope that is preyed upon by this insidious lot of thieves—many people with cancer and other fatal conditions have been promised cures or treatments that are either ineffective, dangerous, or don't even exist and I have considerable compassion for those that are victimized by these schemes—they are simply trying to find a way out. There is also, as you have surely seen by now, no *shortage* of ads promising cures for all manner of sexual "dysfunctions" and offer enhancements that *stretch* the limits of credibility (pun intended, sorry even I need a little crass amusement).

The new "generation" of threats, and a precursor to identity theft, is referred to as "Phishing" and it has become a burgeoning cottage industry for thieves because, sadly, it works. In a Phishing Scam, thieves send e-mails that look identical to an authentic message from E-Bay or a well-known bank that requests very personal information under the pretext of updating your accounts. The e-mail directs consumers to an equally authentic web site that even has a name like members.ebay-services.com (not real). The form then asks increasingly personal questions leading up to a long list of red flags. No organization, that I am aware of, would ask you for your ATM PIN number and bank transfer code. But these do, and people fall for it all the time. I get two or three of these e-mails a week and would be quite surprised if you haven't received them already. The best way to contend with this type of threat is to either delete them or, preferably, report them to the targeted institution.

Though identity theft may not appear germane to the subject at hand, it does however involve the covert collection of personal information and to that end that alone justified its inclusion. I would think that awareness begets vigilance and just knowing these scams exist will provide readers with the information needed to avoid them. Lastly, that brings me to what I would consider one of the most egregious threats that has come to the point of causing irreparable harm. Spy ware, first observed in the late 1990's, has seen explosive growth and represents one of the most treacherous of threats. It is the pinnacle of deceptive behavior and there is no rationale defense that can be made. To quote the definition of treachery is to ultimately define spy ware.

treach·er·y , *noun,* (trĕch'ə-rē)

"\Treach"er*y\, n. [OE. trecher["i]e, trichere, OF. trecherie, tricherie, F. tricherie trickery, from tricher to cheat, to trick, OF. trichier, trechier; probably of Teutonic origin. See Trickery, Trick.] *Violation of allegiance or of faith and confidence; treasonable or perfidious conduct; perfidy; treason.*"

I haven't seen the term "perfidious conduct" in quite some time but it is rather fitting under the circumstances. Spy Ware, to be concise, is software that is installed on your computers, often without your knowledge, for the express purpose of *covertly* gathering information or *overtly* influencing behavior. That was the short version. At the more benign end of the spectrum, Spy Ware can be designed to spawn pop-ups with advertisements that you never asked for. But as you move across the continuum of this class of software, the function of the software becomes much more treacherous.

The majority of Spy Ware programs were designed to capture marketing demographics and behavioral data such as keystrokes, sites that were visited, and other habits that can be used to determine buying behaviors and personal interests. If you had an interest in football, and likely visit football related sites, you could find yourself wondering why you are being deluged with offers for football equipment, tickets and so on. At the other, and most extreme end of the spectrum, where Spy Ware becomes much more malignant, the software can cause serious harm to your computer, steal information, and force behaviors that would otherwise not be allowed. Some applications can alter default home sites, add favorites to the desktop, or capture instant message discussions that you are having with friends. At worse, there is the potential that Spy Ware can be used as a weapon of industrial espionage, taking over your PC, and using it as an anonymous gateway to an intended target.

To compound the problem, there are, according to the Aberdeen Group, approximately *7,000* of these applications floating through the Internet and with that statistic a very high probability that *your* PC is infected and cleaning spyware out of a system can be notoriously challenging because they have designed it to be impossible to delete it. At a structural level, spyware is a small application that is "bundled" with another application to hide its real purpose; for the most part spyware can be found in virtually any software but it is most common in "free" software, games, and even MP3 files. The most visible and egregious cases involved file sharing software such as Kazaa that are bundled with several spyware applications. Though it is beyond the scope of this book I would recommend using several different anti-spyware suites including LavaSoft's Ad-Aware, PepiMK Software's Spybot Search and Destroy, and Microsoft's own anti-spyware suite that you can get for free from http:/www.microsoft.com/security. It is no longer a case of if but rather when you will get hit and knowing how to clean your PC will reduce your target profile.

Since I have devoted an *entire* section to Identity Theft in Chapter Seven, I will keep this discussion relatively brief. Identity Theft, unlike computer viruses, is much more difficult to resolve, and once it happens, the losses to a person can be catastrophic to say the least. Identity thieves don't need the Internet to steal your identity; we already make it too easy as it is. But the Internet begets them an information arsenal that would make the C.I.A. proud. Once you have been targeted, the act of impersonation becomes quite simple and before you know it, someone, pretending to be you, just bought a Mercedes.

When it comes to the threats discussed here, it becomes apparent that users are under siege from an unscrupulous lot that have invaded our homes vis-à-vis through our computers; such acts constitute a violation of our trust and confidence and also do untold damage to the many legitimate companies and sites and play by the rules. Though these threats tend to focus more on your financial and physical security, they all involve a fundamental violation of privacy—thieves still need to understand you to steal from you and the more you can keep from them the less appealing a target you become.

3.4 Protecting Yourself Online

"Common sense is the collection of prejudices acquired by age eighteen."
—Albert Einstein (1879–1955)

This book was never intended to focus on the Internet and there are enough books already devoted on the subject so, with that in mind, I won't spend an inordinate amount of time talking about PC or Internet security. Whatever the case, that so many threats exist online is well known and yet the majority of internet users are blithely unaware of either the dangers or even the simplest of countermeasures that could protect their homes and families. As it was once said, there is a sucker born everyday. Thus giving into temptation, I have decided to cover some of the more commonly known means of protecting yourself. Much of that advice is more *common sense* than technical savvy and it doesn't require a degree in computer science to play safe.

Never at any point in human history have we had such an ability to share and communicate with everyone on the face of the earth but, paradoxically, wherever there are large collections of people, those that seek to cause harm are never far behind—and to them the Internet is one big all-you-eat buffet. In light of the threats I mentioned earlier, the goal is to find that precarious balance between enjoying the online experience without falling into its darker trappings. The more you know, the more vigilant you are, the less likely harm will come your way—it really is that simple. Some of the primary threats are physical and can be reasonably reduced or eliminated by keeping your computer safe from prying eyes; but there are many threats that have nothing to do with viruses or worms and everything to do with keeping a healthy sense of skepticism and having the conviction to say "No". Let's take a look at the easiest of problems and then I will tackle the more subtle threats later.

Protection of your home computer or laptop, especially if you have a cable modem that is always connected, is your first line of defense.

- Investing in up-to-date virus protection is an absolute must given the proliferation of viruses and worms that are now common in e-mail and attachments. No matter the package you chose, it must be current and have the ability to constantly be updated with new virus definitions. It isn't just viruses—your PC can also host Trojan Horses that are usually contracted by downloading music or games of the Net. A Trojan Horse is usually malicious code that is designed to look like something enjoyable—that MP3 you love so much could very well be collecting all e-mail addresses or scanning for credit cards.

- Install a firewall to block the unused ports on the PC. This will also prevent others from using the PC as a jumping point to other PC's. Indeed, some ISP's will go so far as to shut off your service without telling you if they discover your PC is propagating "spam" or if they discover you have been unwittingly hosting a web site. If you are not comfortable with installing your own, your local PC shop or even the big box stores often have staff that are qualified to install it for you. And while I am on this particular topic, it is also considered a best practice to turn off unused computers that are connected via cable or DSL modems; you'll save power and headaches with that simple act.

- Acquire a "disposable", clean e-mail that can be used when web sites demand so-called registrations or if you are required to communicate with those you do not know; I would strongly advise that you *not* use your e-mail address from work or home. You can quickly get such an account with Yahoo, Hotmail, or many other free e-mail services and they are easy to dispose of when the spam becomes intolerable.

- Read the Privacy policies posted to the sites that you visit, but don't believe everything you read. Privacy policies are supposed to inform consumers of the information practices of an organization; but those policies are only as effective as the thought and determination that went into them. Policies can also change very quickly without your knowledge as it did with Amazon.Com and Double Click.

- Think twice before providing any kind of information to a web site, no matter how well known they might be. Remember, that information will be used to profile you and could very easily be sold to third parties. Be equally wary of questionnaires and surveys; and if a site demands information in exchange for something you need, consider looking elsewhere.

- Absolutely, under no circumstances, should you ever reply directly to the junk mail you receive; as soon as you do it validates the e-mail address and it will probably be sold seven times before your first morning coffee. Either block the address or send the e-mail to the abuse coordinators that are emerging in many ISP's. On that note, bear in mind that junk mail costs billions of dollars a year in lost productivity and misuse of corporate assets; if you can identify the identity or target web site of "spammer" then send it directly to the coordinators or security group of your ISP—if your getting one e-mail then it is likely 100,000 identical e-mails came through your providers' mail gateway.

When it comes to other forms of communication on the Internet, be it e-mail or chat, privacy and security is almost all but forgotten and discarded. If you are not concerned with who reads your e-mail then there is nothing to be concerned about, but if on occasion one of those communications is very personal and private, then take a moment to re-evaluate *what* your sending, to *whom* you're sending it, and *how* it is being sent.

For one that has spent many years studying cryptography, what you may know as encryption, and several years of doctoral work dedicated to that field of study, there is a natural bias on my part to question why so few are using encryption to protect data and e-mail. Encryption may provide the greatest defense from prying eyes but even anecdotal evidence shows over 99% of the e-mail crossing the globe is unsecured—in fact most Internet traffic itself is unsecured. Even if an e-mail safely reaches its destination there is still the risk that the mail server itself could potentially be compromised. Kevin Mitnick, one of the more "celebrated" hackers of his time, was convicted for penetrating ISP's and stealing credit card information in plain text files. Though I have been using encrypted e-mail for years, there are so few friends or coworkers that are equipped with the software that I am frequently forced to send e-mail in the "open". In fact with some of those people, given the nature of our discussions, it remains safer just to call them on a "land-line". So as much as it might pain you, a little lain discussion on encryption won't kill you.

Though rooted in theoretical mathematics, the easiest form of encryption to comprehend is Symmetric Encryption where both users use the same key to scramble a message. The simplest of ciphers, known as the Caesar Cipher, adds or subtracts characters. If you take the word MARK then shift each letter two places to the right, you get OCTM—utter gibberish to an interloper unless they know the "key"—the receiver reverses the process and shifts two characters to the left and the original plain text is revealed. Ciphers have been around for thousands of years for a good reason—they work. The problem with this class of algorithms is the difficulty of exchanging a "key" between both people before the message is scrambled and sent. So if I need to send a secure message to a coworker, I must call them first and give them the key. Common algorithms in this class include Blowfish and DES.

The problems associated with the sharing of keys gave rise to the next generation of encryption known as Asymmetric Cryptography that uses a different key to encrypt and decrypt. The mailman would have one key to put the mail in each slot of a building mailbox and you have a different key to open your own box. In the same manner, asymmetric cryptographer works in a similar way. Now I can find your public key in a directory to send you an encrypted message and you possess your own private key to open it. That is much better key sharing. But how do you know a person really send a message? Without some form of validation, I could very easily "spoof" the header of an e-mail to look like it was coming from someone else and cause all sorts of mischief. The answer to that is the Digital Signature. The key you use to open your e-mail can also be used to sign outgoing e-mail—conversely your public key can be used by a receiver to verify your identity. And since you are the only one who knows the pass-phrase, no one can duplicate your (digital) signature. It is impossible to reverse a one—way function.

I have been using PGP for many years and consider it one of the best for small, individual use but there are many products that work just fine. There are also "web based" alternatives such as Hush Mail (http://www.hushmail) that I use when I need to send *very* sensitive data. It is also quite useful if you cannot install software at work or home and the basic service is free. The short of this discourse is simple, if you have something you don't want read by the rest of the world then either find a way to secure it, find a more secure way of communicating it offline, or don't send it at all.

So far I have focused on the threats that you face, but it is even more the case for children given their lack of experience and judgment; I would not want to deprive children from the benefits that the Internet can provide, but parents, guardians, and educators have a responsibility to protect children whether they are in the school yard or in front of a screen. And that can be quite a challenge when many children are more Net-Savvy than their parents.

How you interpret the threat to a child is highly subjective and it might include everything from sexual predators and pornography to highly manipulative advertising practices. While I cannot stress enough the need for adult supervision, an adult is not always going to be present so it is equally important that safety be taught to children as early as possible. Here are a few of the most common guidelines for children that will serve as focal point for your discussions.

- Children must realize that things are not always as they appear and that they must learn to never give out personal information to strangers whether they are online or not; that might sound terribly cliché, but a child may not perceive the same risks surfing the web as they might perceive in a park. In the park, the threat of a stranger is much more tangible—a so-called friend chatting online doesn't look like a threat.

- Related to the above, try to get familiar with who your children e-mail and chat with online. If possible learn how to read the logs (i.e. mIRC) and watch for unusual behavior like excessive amounts of time spent at their computer.

- Remember that your child's friend might not be the little 12 years old down the street, but a 40-year old repeat offender across town. That is an unfortunately negative view of the world but I would sooner put that thought across your mind than have to read about it in tomorrow's paper. Children have and will continue to go missing because of precisely this kind of threat.

There is no such thing as a perfect technology and it is very unlikely that it will ever exist. To paraphrase a colleague's e-mail tag, humans are not perfect so why would we expect our technology to be perfect—the Internet is no exception to this truism. Even the best filter software will inadvertently let malicious web sites in while blocking legitimate sites; don't rely exclusively on technology to parent your children or protect yourself. Technology can, to some extent, ease the burden of providing a defensive perimeter but it is always shadowed by personal vigilance.

I use the term vigilance quite frequently because it is the most important weapon that you have. It is a fitting closure to this chapter to emphasize the need for a little common sense and self-control in the face of an unrelenting torrent of offers, ads, and all things free. Likewise, the enforcement of policies and the law is only as effective our willingness to take the time to report improper behaviors. An Internet Provider cannot control junk mail unless we report it and the police cannot prevent frauds and other crimes unless we make the effort to report them.

3.5 The Best Defense Is Common Sense

"I happen to feel that the degree of a person's intelligence is directly reflected by the number of conflicting attitudes she can bring to bear on the same topic."

—Lisa Alther, Kinflicks, 1975

The Internet, despite its excesses and limitations, is nonetheless one of the most intriguing studies in technological and cultural transformation. and there is little reason to expect the pace of change and progress will decelerate soon—on the contrary, though presumptuous, I would expect the pace of change to accelerate even more so within the next ten years with an explosion of users, the introduction of new technologies, other cultural shifts that will place more emphasis on the digital persona.

And accompanying this cultural revolution will be a renewed focus and debate on everything from fundamental freedoms to expressions of the self, self-determination, autonomy, and of course privacy. In my opinion, the battle over privacy that we have witnessed thus far will pale in comparison to the battle that will be waged over the next few years and the Internet, while only one technology amongst many, will play a role as subject and instigator.

The Internet has proven to be a lightning rod not only for those that decry the rise in gambling, pornography, and other vices but also for those that do not share an appreciation for freedom and self-expression, especially when those "self-expressions" clash with their belief systems; and it is impossible to be thankful for the freedoms held in North America compared to countries that are not so enlightened—there are still many Countries such as China where the controlled from the top down and expressing one's thoughts can still lead to a lengthy prison term.

From the perspective of privacy on the Internet, here too there is a never ending conflict between your desire to remain anonymous and those factions that do not adhere to this principle; just as society has gradually moved from an anonymous cash based culture to an identifiable card-centric society that has resulted in transaction trails, the Internet could potentially move towards an identity-based infrastructure. Truth be told, it isn't that difficult to determine a person's identity from their IP (Internet Protocol) address and it is becoming increasingly more arduous to remain even remotely pseudo-anonymous. For those that are technically savvy and have "lived" on the net for many years, cloaking and proxy surfing is part of our natural lexicon.

As it was articulated in Section 3.2, the Internet was not always like this but then again the Internet of the 1970's was not the same beast it is now; the threat of nuclear war was is all but a distant blur of history and has been replaced by different concerns and threats such as terrorism. This, however, is the 21st century and times have changed—nuclear war has been replaced by terrorism, transnational corporations have been replaced by global outsourcing, and Watergate has been replaced by the trials and tribulations of Martha Stewart. On closer examination the events of times past may have been replaced by more contemporary events but the underlying themes of ideology, greed, arrogance, and pettiness have not changed at all. And even though over 40 years have passed since ARPA was formed, the very same themes are self-evident wherever you go on the Net.

This global network, regardless of its greatness, is still just that, a network—and a network is technically nothing more than a motley mixture of routers, switches, and hosts. What makes this phenomenon so interesting is not the hardware or software; it is how it is applied to solve (or create) problems. It is often lost on technocrats, but technology by itself is useless—a creative, oversized, metal paperweight. If people did not find the means the leverage on this technology, it would have remained an academic curiosity or a relic of a distant era.

In spite of all the threats and dangers that have been enumerated, the Internet is still a marvel both in terms of the technology that drives it and the manner in which people have learned to leverage it. Culture has indeed changed as a result of the Internet and yes, there are downsides that must be openly confronted but it has brought forth positive changes as well that are reshaping entire industries.

The act of book publishing exemplifies the Internet Generation. In times past, the pathway to getting published was a long and strenuous journey and fraught with uncertainty and the subject tastes of editors (it still is). Now authors can write, publish, and distribute their literary works through cybernetic publishing entities that offer speed and flexibility that are simply not found in traditional publishing houses. It does not mean a book will do well but at the very least it provides an opportunity for an author to compete in that market. I may never find wealth from my worth but I will find it on Amazon much quicker than if I would have taken a more traditional approach.

The same can be said of many pursuits. Musicians that couldn't get the time of day from a recording label can go "Indie" and publish their own music—the proliferation of file sharing and ensuing lawsuits is one instance in which an industry is grappling with the vision of its own extinction. Either the music industry will adapt...or it will perish. Similarly artists that would not otherwise be given gallery space can, by merely setting up their own site, exhibit their works of art. And Entrepreneurs can challenge traditional bricks-and-mortar companies in a free market Internet economy where previous traditional barriers would have prevented their entry into the market. But a presence on the Net does not automatically beget success and it is the market and the public that will ultimately determine who is or who isn't successful. You can have a beautiful web site and *still* sing worse than those that have disgraced themselves on American Idol. The Internet simply offers a substrate for human creativity and expression even if the quality of such creativity is open to subjective interpretation.

Sadly, the things that have made the Internet so attractive to millions of people have acted as a magnet to those lacking a moral compass and one of the costs of this new found freedom are the threats that come with it. In the early to mid 1990's it was considered a privilege in most homes to have a PC and the thought of connecting to other parts of the world was known to a very select few groups during that time. Now, Internet access is more the norm than the exception but it has brought with it the deluge of spam and spyware.

You certainly cannot blame anyone, after reading of these threats, if they were to unplug their computer and toss it through their window; but that was not the intent of this discussion nor is it the solution to the problem. Insofar as improving your security, the solution involves quite simply paying more attention and developing a sense of situational awareness. That entails taking precautions to prevent the physical threats I discussed and keeping a clear mind when it comes to the more subtle of threats. The best I can hope for at this point is that you will take the time to review the recommendations I've made and at the very least keep your systems up to date—that won't solve all the problems but it will eliminate a majority of them.

In Section 3.4 several protective measures were mentioned, some obvious and some that may be new to you. At a technical level, it means protecting your computer against the most common types of threats such as viruses, worms, and Trojan Horses—which is simple enough if you make the effort to install the most up-to-date anti-virus software you can find. The same is true of installing a firewall to prevent the unauthorized use of your PC without your knowledge. This alone is going to save you considerable grief and possibly prevent the loss of data or theft of information. That is worth the chapter in itself.

Despite my best efforts, my experience tells me that a majority of readers will not make a move to encryption any time soon and in all fairness it is still complicated for those that not technically inclined. Hopefully that will change over time and I would like to see Microsoft provide that functionality to make it easier for home users to secure their own communications. It ought to be built right into Outlook. The other methods have nothing to do with technology and everything to do with our ability think logically and assess situations clearly and without distraction. It requires vigilance, commitment, and self-control not to be taken in by promises of wealth or happiness when these represent the some of the most fundamental beliefs of the human species.

To put computer security in a proper perspective the best place to bring this discussion to an end isn't through metaphors but experience. In January 2005, while enduring yet another edit of this book, I was working on my laptop in the strangest of places—downtown and sitting in my truck. Within less than five minutes I had 1) discovered 12 wireless networks within a three block radius 2) determined 65% of these wireless networks were unsecured 3) acquired an IP address 4) surfed the net from my truck and, most disconcerting, 5) was able to easily browse through some of the unsecured hosts. And when I say hosts I really mean *homes*. And that was without any effort on my part. Had I possessed a malicious disposition it would have been "child's play" to either destroy their systems or steal all manner of documents. It would only have taken those users a little bit of reading and a few minutes to secure their wireless network.

It is unfortunate that the threats I've spoken of and the relentless media attention to the negatives tends overcast the greatness that the Internet can offer when used for the betterment of others. I keep thinking of the many support group web sites and chat rooms for the terminally ill, the efforts of government to improve the delivery of electronic services that offer convenience, and the ability to research virtually any topic you desire. There is much good to be found on the Internet, and I would sincerely guard against the erroneous perception that the Internet is little more than a cesspool of pornography and file sharing. Think of all the discoveries in science that are being accelerated through improved collaboration by universities connected in real time around the world. Think of remote villages in North Africa that are now connected to the Internet that for the first time allow its residents to see a world they never even knew existed. In other words, its easy to be negative and find fault; it is much more rewarding to find promise than problems.

In a nutshell, for all the wonders of technology, the best defense against the evils that await you is just common sense. That is horribly anticlimactic but it speaks volumes of this silly expectation that technology fixes everything and underscores the unwillingness to take responsibility for our own actions. Chapter Three offers you the arsenal to guard against those that seek to do harm to you but it is up to you to decide what countermeasures you are going to use. That is something I cannot do for anyone—my duty to my readers is to inform and provide the best practices for self-protection.

4

Advanced Technologies

4.1 Advanced Technologies

"Technology is dominated by two types of people: those who understand what they do not manage, and those who manage what they do not understand."

—Putt's Law

fsWhile the Internet still basks in the spotlight there are many other technologies that are equally important but far less visible and which have now or will have a significant impact on privacy. It is too bad that the only exposure the public gets are 30 second sound bites on CNN because, though an improvement over no coverage at all, they still do not adequately describe how these technologies work or what they mean to you. Still it is encouraging to note that the media have been increasingly discussing genetic monitoring, privacy issues and I am pleased to see it is has reached the mainstream. And in an effort to build on this momentum, I've dedicated a chapter to the broader view of science and a snapshot of a few key technologies that illustrate the need for a wider focus on science and technology.

I've also noticed, that in raising the subjects noted above, that even in the absence of a science education, the general reaction tends to be "They can do that?" followed immediately by "Well that sucks"—or colorful metaphors to that effect. While a majority of the public does not know what Radio Frequency Identification (RFID) is, they do understand the basic idea of a transmitter tiny enough they can't see—and I am always pleasantly surprised by how quickly people realize the potentials for good and evil. There is, in this, perhaps a lesson for policy makers too—don't underestimate the intelligence of the public, they do not need a doctorate in Physics to appreciate the importance of RFID or anything else they might try to spin. But first let's step back a bit and look at how technology affects society before getting into how it works.

Significant technological changes have historically resulted in extensive social shifts and looking to our past, particularly the Industrial Revolution can be quite illuminating. Put simply, during this period, large segments of the population were displaced from their lands to search for work that could only be found in the cities that acted as a catalyst for social deterioration, fragmentation of the family, and a general sense of uncertainty. And just as the 1980's will be remembered as a period of financial excesses, the Industrial Revolution displayed similar symptoms and the effect on the population was even more pronounced due to the shift from an agricultural economy to a manufacturing economy and the population that was ill-equipped to deal with these changes.

On the surface, at least economically speaking, all was well and good and money flowed to the merchant classes that were generated by the labors of those not so fortunate; children were forced to work in factories and this "revolution" was a pivotal point in history. For this period of economic revolution birthed the alienation of the working class that has forever changed the relationship between the so-called haves and have-nots. Suffice to say, it is no coincidence that many of the contemporary social problems such as suicide, divorce, domestic violence became more prevalent during this period. Students are often taught that these years were a remarkable time of innovation and growth, and it was, but there is a darker cost that is often conveniently left out of the lecture. That does not diminish the advances of the time, but it does remind us that there are hidden costs to progress that are often ignored or marginalized.

Furthermore, many of the social changes experienced during this period stemmed in part from the development of new technologies and to that end there is a misunderstanding when it comes to why technology progresses and who it was meant to benefit—the altruistic side of us would like to think that technology should benefit society, and perhaps that comes over the course of time, but that is not always the case—in fact technology is rarely if ever shared equitably with society. To head off the calls of "Down with the rich", it is important to note that without the investment capital provided by the proverbial merchant class, many of the conveniences that exist now would have never come into existence. And for that reason, investors cannot be faulted for expecting either a direct financial return or a competitive use of a new technology to mitigate the risk of failure.

Still, the fact remains that many new advances are specifically designed to eliminate the human labor component and specialist knowledge and a visit to any modern factory and the proliferation of robotics is evidence of the underlying motivations behind many advances. The same may be said of the 21st century and the growth of global outsourcing is similar to the displacement witnessed during the 19th century. How many people have been "streamlined", "right-sized", and "outsourced" by information technologies that have rendered their functions obsolete? It is not always a direct technological takeover; many North

American's that have lost technical support jobs did so because the labor was cheaper overseas and new advances in telecommunications allowed organizations to integrate geographically dispersed call centers more effectively.

That said, virtually every technology can be classified according to its usage and application and this system of classification can offer clues as to who will benefit the most from its adoption. Stoplights and stop signs, though low-tech can be classified as both utilitarian and dominionstic meaning that while both act to modify human behavior they do so to save lives. Other technologies like biometrics are also dominionstic but generally protect the organization from the individual. Contrarily, a sailboat (for lack of a better example) is utilitarian since it is defined as a mode of transportation but can be classified for pleasure. There is absolutely nothing to prevent any given technology from being associated with more than one class. Going to a movie theatre falls within an economic class because it benefits to the owners but it also provides pleasure to the consumer. There are quite a few classes including those that I have mentioned as well as scientific, esthetic, symbolic, and linguistic.

The technologies I present here are by no means an exhaustive study but they do capture some of the most important advances and provide a sufficient snapshot of scientific progress to date. Similarly, in one way or another, these advances either have or will eventually permeate the mainstream.

The first technology, data warehousing and data mining, is the most prevalent but least visible and deserves a berth of its own given the magnitude of data collection and, of course, the fact that it is generally invisible to the public and thus given little thought. Still, information is collected on every one of us and the appetite for information grows more intense by the day. Smart cards were the second choice since everyone has at least one piece of plastic in their wallet—not just credit cards but loyalty cards, social insurance cards, drivers licenses, and variations on that theme—if you have any plastic, and that card identifies who you are, then smart cards present a complex balance between necessity and convenience and a potential threat to your privacy and freedoms.

At the more advanced end of the spectrum, I wanted to include a segment dedicated to RFID since you may, sooner or later, be confronted by it. These tiny, grain-sized, chips are finding their way into everything from clothing to cans and have the potential to allow merchants to track your movements through the store and whatever you may be carrying. The final two segments are dedicated to technologies that are even more advanced—location tracking and genetic monitoring.

A lot of the material in this chapter can be rather intensive and it might even take a little longer to get through it but that was a necessary evil. Nevertheless, in reading each segment try to bear in mind that every technology, barring a few cases, is capable of great promise and great harm, depending on how it is to be used. There are very few advances in science that I would want to ban entirely and it is possible that, in banning a scientific line on inquiry, we may be causing more harm than good. So keep an open mind while you work your way through the chapter.

Naturally, everyone is entitled to their opinion as to how technology affects society and those opinions occupy the spectrum of beliefs ranging from the unjustly pessimistic view of the disbelievers to the wildly rosy predictions of the purveyors of change that profit from its proliferation. As well attitudes towards technology are affected by scientists, politicians, policy makers, and special interest groups that have their own agenda. Obviously some changes will be positive or negative, direct or indirect, and so on.

In order to properly evaluate the impact of any innovation first it must be determined what the technology does, the impact on society (be it positive or negative) and who will benefit from the change. To illustrate a more balanced assessment of social change, consider a few classifications I created for this discussion.

- Those that possess a direct, positive benefit to society such as traffic lights or the development of new flame-resistant materials—both of these innovations have a direct impact on the quality of life and the ability to save lives. The positive element was a component of the design methodology. From a privacy perspective, this may include easier-to-use encryption software.

- Those that possess an indirect, positive benefit. Online banking provides direct benefits to the bank, not the customer, in the form of reduced labor and improved efficiencies; Customer's do, however, enjoy new conveniences and these should not be understated. In this classification, privacy would be improved through improved security measures that were designed to protect personal information.

- Those that possess a marginal or negligible impact on the public and have no impact on privacy; modifications to manufacturing or mining processes would have no direct impact on individuals but serve to improve the economic efficiencies of the stakeholders.

- Those that possess a potentially negative impact on society or an individual and which may also have a negative impact on privacy. Biometrics, data mining, and genetic testing each has the potential to bring positive changes to people's lives but they also carry significant privacy risks that have not been resolved.

There are more formal, rigorous models but this suffices as an informal means of more objectively evaluating what you see on CNN. Quite frankly, just asking whether a new technology will help you or harm you is far better than the apathy that dominates the collective attitude towards science. Since, however, this book is about privacy, it is this fourth class of technological changes that is relevant to privacy since those that are classified in this group have the dual potential to provide significant social benefits and yet possess the potential for abuse. The burgeoning market for computers and telecommunications raises a lot of questions and it can be difficult to classify a technology into one group. On the one hand the evolution of computers and cell phones has given society unparalleled access to the world and has virtually eliminated borders. And yet the very same technology has changed society irreversibly—sometimes for the better and sometimes not.

The high costs of development coupled with the pressures of competition do not provide organizations the luxury of developing or adopting technologies unless it will provide direct benefits in the form of reduced costs, optimized processes, or improved market position. There is a dual motivation at play that combines the desire to profit from such discoveries while also improving quality of life. Technological changes can also have a dramatic impact on unrelated markets and computers are a perfect case of a new technology that was grossly underestimated both for its technological and social value. Though it was quite evident that competing mechanical devices would eventually lose ground to computers, it fails to account for the ways computing has changed everyone's lives and the genesis of a completely new industry that has brought with it new opportunities both for individuals and organizations.

Then again I am not completely convinced that we are better off being wired to the extent that we are now and I believe it presents, metaphorically, a double edged sword. On the one hand, it is difficult to envision life without computers or cell phones and is impossible to fathom a world without them given the dependence on technology. How many channels do I need for my TV? I have trouble finding something to watch on the 58 channels I have now let alone the 500 channels they are offering? How many key combinations can they possibly fit into a single cell phone? How many remotes do you need to control the TV, VCR, DVD, and Satellite? Despite whatever advances have emerged, there are compelling statistics that this omnipresent state of connectivity is beginning to have an adverse impact on our physical and emotional health that were not as prevalent 20 years ago—most notably the expectation that people are available 24/7 and the encroachment upon private time that has become the norm. This problem has become so acute that some companies have essentially had to "unplug" their employees.

To switch gears a bit lets refocus the discussion on the impact of technology and one of the more contentious subjects of the 21st century. Due to my personal interest in medicine and molecular biology, I tend to use medicine as an example because the impact of ones health is easier to articulate than the abstractions that are required to grasp the potential threats of data collection. Contrarily, it is very easy to grasp the effects of illness and the impact of medical science. I've touched on some thorny issues in earlier chapters but embryonic stem cell research is definitely one of my favorites.

Here is the short version. In the simplest of terms, stem cells can be found in bone marrow, in the blood of the umbilical cord, and other parts of the human body such as the brain. These cells have already been used in transplants for years, but it has only been in the last few years that scientists discovered that stem cells can evolve into other types of cells or for that matter into any of 210 types of human tissue—they are the equivalent of a manufacturing facility. Stem cells can be extracted and inserted into the embryos and in turn the embryos code the stem cell to develop into target cell types. There is still much that is unknown and the research is in the early stages but the potential is staggering.

Naturally, the research demands a supply of human embryonic stem cells and that is where the battle lines have been drawn. That battle stems from the belief in some circles that the destruction of an embryo constitutes a form of abortion even though the federal government has gone out of its way to avoid the term and have attempted, as politicians are inclined to do, walk both sides of the fence by allowing a very restricted number of stem cell "lines" that will fail to pacify either faction. The scientists want unfettered access to all lines and religious groups are lobbying for a complete ban on stem cell research. This is a no-win situation.

The underlying ethical question is whether the ends justify the means. Even if the objective is to eliminate death and suffering, if you hold to the belief that this constitutes abortion, then it is unlikely you will be swayed by the merits of the science or the potential to save millions of lives. Of course the argument can be reversed too—what is one embryo when compared to a million lives or even 20 lives. There are no equations to ease this moral dilemma and I don't think we have seen the end of this fight. In my opinion it will continue whether extremists succeed in blocking U.S. research or not. Even if they are successful in banning this research, laboratories all over the world are already conducting this work and they will not hesitate to fill the void left agape by the United States. The United States could well find itself struggling to maintain a competitive edge in the bio-technology field as it struggles simultaneously with the cultural divisions that threaten its progress. Britain has already permitted the first tests to clone a human embryo and others will surely follow suit. To use a familiar expression, we have already opened Pandora's box and the best we can hope to achieve is to agree upon an ethical standard for biological research. More often than not contentious discoveries have historically resulted in social divisions and stem cell research is one of the more graphic cases in which technology is moving faster than our ability to come to terms with its impact.

Several lessons and observations can be drawn from this single ethical dilemma such as the trade off between means and ends, the common good versus the individual desire for survival and these questions are not the exclusive domain of doctors and politicians; whatever decisions arise from this debate could very well have an impact on how long we live. In other words, it is important to you because someone, with an agenda that may not be yours, is making decisions that involve you. For patients suffering from all manner of diseases, this debate is anything but academic.

In the end, the social changes that result from technological innovations are just as if not more important than the technology itself and more attention must be given to the moral and ethical questions that must be addressed and which frequently pit economic values against myriad social values. Assume for the moment that a new drug is discovered that would cure every cancer patient in the world but to complicate matters, the cost of developing the drug would cost $20 Trillion. Who is going to pay for the research and development? Who is going to own the rights to the drug? Should a government intervene and seize the drug for the common good? Should the pharmaceutical company not be allowed to recoup their costs? Should the patient have to pay for a drug that will cost them their home or life savings? Too many questions and not enough answers.

Similar questions arise when there is an insufficient market to develop a drug that might save the lives of a small number of the population, especially when the number of patients would be insufficient to be profitable. Would you want to be the one to make that decision? How does an executive, assuming they would ever have the courage, tell a patient or their family that they are not "economically viable"? These are melodramatic instances but they do articulate the moral and ethical challenges that arise with the discovery of new technologies. There is so much more to a technology than the code or components that make it work and I believe more attention must be dedicated to the social impact of technology before it is unleashed on an unsuspecting populace. That, sadly, is wishful thinking.

4.2 Data Mining

"Data is not information, Information is not knowledge, Knowledge is not understanding, Understanding is not wisdom."

—Cliff Stoll & Gary Schubert

Barring those in the computer field, there are few that would characterize databases and data warehousing technologies as interesting let alone appealing compared to other technologies like biometrics or RFID that seem to get all of the attention. Nonetheless, it is the engine upon which everything else is based and we are remarkably dependent on this unsung hero champion of technology. In fact it is safe to say that virtually every organization on the face of the planet has some form of database platform for managing the massive amounts of information that is collected. Further, it provides the storage and structure necessary for the development of so many other technologies and it stands, on its own, as one of the technological pillars of society.

The privacy issues stemming from the explosive growth of databases make this the first topic that I would like to explore with you. It is not just the size of the databases, it is the ease with which heterogeneous platforms can be connected, who has access to the data, and how that data is secured that have an impact on everyone. Even if you are a technophobe data is still being collected about you and stored in a database, indeed multiple databases—every government agency and every company you have ever dealt with has a "file" on you.

For those not familiar with databases, a database is nothing more than a means of storing data in a way that makes it easy to store, locate, and retrieve. As a former senior trainer that spent many years teaching database design and architecture to large organizations, teaching these organizations how to connect even more databases together is a component of the curriculum and one of the most requested topics in those courses. Likewise, in the course of teaching relational theory, a favorite module in the curriculum was the joining of tables that provided larger aggregate views of data from multiple sources. Though security is covered, privacy most certainly is not.

Without getting into a discussion on the relational theory of heap spaces and outer join predicates, our digital personas are essentially a series of data elements that are indexed on a unique identifier such as an invoice number, credit card transaction, drivers license or whatever comes to mind. To give you an idea of the underlying problem, how hard would it be to connect your driver's license or social security number to your credit card number and your credit card number to other repositories such as traffic tickets, or the medications that have been prescribed? At what point do these burgeoning repositories and their keepers turn to something more unique like fingerprint?

Even when databases are properly designed and deployed, there must be a means of transforming trillions of meaningless bytes (therein referred to as *data*) into meaningful *information* on which decisions are based. There are many different types of extraction queries though most tend to fall in the category of transaction-centric questions such as "How much did we sell?" or "How many rooms did we rent out?"; these questions are useful but they tend to represent a snapshot in time. A more sophisticated family of technologies use very advanced mathematic methods like K-Nearest Neighbor to answer complex questions involving patterns and to discover relationships in the data that were previously unknown. Data mining is important to privacy because it allows organizations to sift through tons of data to discover patterns about us that the organization was not aware of. As these techniques and technologies becomes increasingly automated and intelligent "they" will be empowered to make decisions autonomously that have the potential to affect individuals.

The primary difference between traditional analytical queries and the new generation of autonomous discovery is that the relationships in the former are understood compared to a lack of assumptions in the latter. In the case of the latter, algorithms are essentially freed to *find* correlations. What makes this class of algorithms all the more fascinating? These algorithms actually have their rootsin Artificial Intelligence. What was once a simple SELECT query running against a 1 Mb Access Database is now a 7 Tb (Terabyte = Trillion) data center using genetic algorithms, evolutionary algorithms, rule induction, and fuzzy logic.

Walmart, for example, is reported to have the largest and most advanced computing infrastructure in the retail world, perhaps even in the entire private sector, and you can imagine how complex their questions are. Rather than asking how many boxes of soap were sold, the "big" thing is pattern recognition. A pattern recognition algorithm may be able to determine that 78% of customers that bought one variety of cereal bought 2% Milk while 17% preferred homogenized milk but only if it was with corn flakes and not some other cereal. It might even determine the demographic grouping, optimized shelf distribution, and time series analysis. Genetic algorithms, which I am particularly familiar with, could be used to optimize which boxes go on which shelves based on sales, customer preferences, behavior studies, and rule sets. It is analogous to an ultra-fast game of Tic-Tac-Toe except each "square" represents a shelf location and an aisle. These are rather benign cases but the potential exists to use these technologies for more sinister purposes.

Prior to the use of these algorithms, data often comes into a database in varying degrees of completeness, accuracy, and consistency which brings me back to data warehouses that play a fundamental role in this equation. These massive repositories serve as the central clearing house where information from many sources is pulled together, cleaned, and organized before the analytics are applied to the data. Though hardly exciting stuff, some of the core functions also include security, encryption, indexing and optimization, storage management, and disaster recovery. It gets far more complicated that what I am providing in this book but it may also require replication to other sites or geographic locations, indexing, and transformation services that transform inbound data into a form that can be searched more efficiently.

By comparison, data mining relies on a discovery-centric approach that uses everything from genetic algorithms to neural networks and pattern matching equations; this is not your everyday database. Rather than knowing precisely what to look for, data mining takes the approach of not knowing in advance what the relationships will be and does not make assumptions about how the data is structured or related. A bank, for example, might use a neural network to determine what types of transactions constitute a possible fraud based on historical patterns or the same algorithm(s) may be used to determine which customers might be interested in other services based on their banking practices and historical transactional behavior.

What is driving this push towards massive database technologies? Most companies do not pursue technology for the sake of technology; Companies, more often not, do not pursue technology for the sake of technology; there must be compelling competitive benefits to adopting a new technology given the often significant costs and risks incurred with the development of these systems. Data warehouses and data mining have, thus far, been known for a large number of project failures—the science is sound, but many organizations do not know a priori what is expected of these systems and that can result in reduced expectations or possibly the perception that it was a failure if it doesn't deliver. So it isn't the technology that's pushing the curve.

A more plausible reason is the increasingly competitive marketplace, saturated by competitors and a fickly consumer that makes acquiring customers more expensive and more difficult than ever before. This eventually led to the de-massification of industries and towards personalized, individualistic target marketing that was originally termed by Alvin Toffler and which began during the industrial revolution. The information age exponentially increased the competitive nature of the market and has in turn resulted in an insatiable appetite for information.

By the mid-1980's to 1990's it also became apparent that there was untapped gold to be found in existing databases and the ability to link isolated ultra-large databases began to mature. Other developments including the Internet, single customer views, and reductions in the cost of storage made the pursuit of large data mining projects more economical and were further justified by the need to compete with those engaged in similar pursuits.

Applications in data mining are virtually infinite and it would be unfair to characterize the technology as evil. To use Walmart again as an example, the company captures point of sale data from over 2,900 stores in six different countries that is eventually stored on a very enormous 7.5 terabyte (trillion bytes) information data warehouse. In turn, Walmart allows over 3,500 vendors and suppliers to access this data to identify purchasing behavior and buying patterns that can be grouped by any type of demographic pattern imaginable. Using similar methods, larger "big box" chains can use extensive databases that are cross-linked with external sources to perform market basket analysis which can also combine customer demographics like age, sex, and geography; such analysis could reveal that men between the age of 21-34 will buy certain items but only during specific times or locations or in combination with other items. These patterns allow retailers to fine tune promotional offers and restructure inventories and store layouts that improve revenues and customer retention.

Governments can also employ data warehouses of similar size and complexity to determine shifts in demographics that alter the need for services. For example, a city may combine tax assessments, health information, and census data to determine which parts of the city are changing over time and how they will have to optimize service delivery to different segments of a community. In my own area, new residential developments have resulted in an influx of a younger, more affluent subset of the population that will eventually require a different set of services. Sudden "bursts" in population can affect housing, schools, policing, and municipal operations and data warehouses can assist in the forecasting of these shifts before they occur.

The privacy issues that are raised by large scale data mining focus primarily on the principles of personal control that govern how information, especially personal information, is collected, stored, secured, discarded, and exchanged. These principles are encoded in the 1980 Organization for Economic Co-operation and Development (OECD) guidelines that form the foundation of contemporary privacy guidelines and statutory law. Ultimately, the risk to privacy revolves around the ability of organizations to use these large database systems to build detailed digital transaction profiles of customers and anyone it wishes. In so doing, someone with a good credit rating might receive a less than favorable loan rate if it is discovered they missed a payment on their insurance premiums. Perhaps, at some point we'll get refused for a loan because, due to such relationships, it is discovered a person used to smoke, thus reducing the probability of long-term survival—it may even be construed that risky behaviors such as smoking or drinking in turn constitute a credit risk. Perhaps it will even become possible for insurance companies to link risk assessment with debit transaction histories.

Data mining, if I was to apply the OECD principles, suffers from several privacy problems that have not been properly addressed. These principles include (a) Data Quality (b) Collection Limitation (c) Purpose Specification (d) Use Limitation (e) Data Security (f) Openness, (g) Accountability and (h) Personal Control and Participation. Though not speaking to every organization, my experience in the past 10 years has been that a majority of organizations do not even remotely know about the OECD guidelines let alone apply a set of acceptable privacy standards as those noted here. While these eight principles are the bedrock of fair information practices it still remains to be seen how many governments and organizations have really put them into practice. Just in the realm of Data Security, the sheer lack of cryptographic usage in even the largest companies does not bode well for the confidentiality of personal information.

Even if the company is protected by layers of external firewalls and encrypted tunnels, the data itself is often raw and exposed. Furthermore, the idea of collection limitations (or scope) and use limitation would defeat the purpose of data mining where the purpose of the data may not be known until the relationships are discovered so it presents a paradox—how do you limit the use of data if you don't know what data you are using. The collection of data is a different matter and is already the subject of legislation that will eventually demand consent from customers prior to its collection; this may put a significant damper on the type of data that is collected.

My primary concern, and it is a concern shared both by privacy advocacy groups, the ACLU, and government privacy watchdogs, is the unauthorized exchange or sale of personal information to third parties that were not directly involved in the original transaction with the consumer; we are talking about the sale of marketing lists and demographic data which contain clearly marked identifiers (name, address, phone) and for which no consent was given by the customer. It may have been buried in a policy document but it is safe to say it isn't posted at the check out counter. I am not opposed to providing *blinded* demographic data (akin to statistics like 21% of consumers bought chips with coke) but the sale of personal identifiers is a violation of consumer trust and many organizations have felt it was "their" data to do with as they please. That, as the law clearly states in Canada and Europe, is *not* the case, and organizations, both public and private will have to come to terms with this new era of protections. The question is, who really owns the data? A company may own the infrastructure in which the data is stored, but they do not own the rights to an individual's namesake.

Databases represent a fundamental technology that will only continue to grow in size and complexity as the appetite for information continues well into the 21st century and there is every expectation that this pattern will continue for decades. The challenge is to allow businesses to conduct their affairs in the interest of competition while ensuring these technologies are not used for discriminatory pricing practices, violations of privacy, and similar offences. It is also just as vital that the data be secured and that organizations adopt fair information practice. To do that, however, first requires management truly grasp privacy and champion that through the company. It is one thing to sing the praises of security and privacy and still another to come to terms with that when I.T. says it will cost 23% more to implement the controls required for confidentiality, encryption, and data retention. It will be even more difficult to carry that note when confronted with the reality of having to notify every customer every time they want to disclose data to third parties.

4.3 Smart Cards

"For a list of all the ways technology has failed to improve the quality of life, please press three."

—Alice Kahn

Whether it be credit cards, debit cards, or convenience cards, it is fair to assume that everyone has at least one card in their wallet and have doubtfully given any thought to the privacy considerations that arise from their usage; but so-called smart cards can carry an abundance of personal information and the more these cards are used, the more they add to the transaction trail that is left behind in the wake of our purchases. What is not always understood is that the completion of a transaction at the convenience store doesn't stop there and the flow of information continues long after a person has left the store.

Needless to say every smart card obviously must provide for a means of authentication because the unique identifier is required so that card transactions are properly applied against the cardholders account. Not every type of card requires authentication and the introduction of convenience and loyalty card schemes have been immensely popular. There are countless variations but the most common are pre-paid phone cards, gas cards, gift cards—it doesn't matter who holds the card since the amount was pre-paid. Add to that the money loyalty cards that provide incentives and discounts. In these cases transactions are debited or credited against the card instead of the cardholder. There are even hybrid schemes in which the ownership of the card is classified as pseudo-anonymous—there is no direct relationship between the card and the holder without cross-referencing the account; nothing is stored directly on the card. There are a lot of benefits to pseudo-anonymous card schemas and I am strongly in favor of them though I am not convinced that the marketplace understands the functionality or value of pseudo-anonymity.

There are many ways to protect the information stored on smart cards not the least of which is encryption that I tend to prefer and the new generation of chip-based cards have the potential to substantially increase privacy or they can also contribute to the invasion of privacy too—it depends on how the card is designed as well as the realization that the more information stored on the card increases the vulnerability of the scheme.

At one end of the spectrum, it is technically possible to store all of the information associated with a cardholder in addition to the transaction data—something we may refer to as a full storage card. However, there are very few card schemes that involve the storage of data exclusively on a card carried by an individual. In general, this would create great fragility in the system, because of the danger of loss of the data without the ability to recover it. Most card schemas use different storage and algorithmic approaches that split the data between the card and the processing facilities.

As a hypothetical example of a pseudo-anonymous card, consider a consumer that signs up for a fictitious card called PSEUDO which allows consumers to transfer a fixed cash amount to the card via debit card, credit card, or wire transfers. Two separate databases are used to maintain the cardholder information and are encrypted using a one-way hash algorithm. Suffice to say, merchants wouldn't stand a chance of reverse engineering the card but law enforcement could retrieve the cardholder's identity through a court order; consumers win by virtue of the privacy embedded in the card the law enforcement win by virtue of their ability to seize transaction histories through the courts.

There are also means of splitting a card so that it might hold multiple applications by using segregation zones on the card that would in turn allow multiple vendors on a single card as you would have with convenience cards. Most likely to be found amongst loyalty scheme cards, there is a serious risk that even with this segregation that transaction information will be shared between the different vendors on the card. A shoe retailer might exchange transactional data with a clothier to determine the likelihood of cross-promotions and offers or just to observe buying behaviors in a geographic area or amongst the stores promoting the card. What do banks and merchants actually see on the cards? It depends on the design of the card scheme but it can range from nothing to everything and further depend on the levels of access. A merchant is unlikely to have access to anything on the card—and so they shouldn't. Even bank credit cards generally segregate transactional details such that the only visible information is when the transaction took place, where the card was used, the amount of the purchase, and likely the status such as approved or declined. Other linked databases would hold exactly what it was that was purchased but access to this information is going to very limited and controlled.

Slowly, ever so slowly, there is some movement towards biometric smart cards that would require a fingerprint (or some other measurement) to complete a transaction. Not surprisingly there has also been a lot of resistance to this concept on the grounds it would be privacy-invasive buthaving worked in this area, there are ways to make this a very secure scheme. Most of the concerns focus on the need to store the biometric template on the card but there are many methods such as split key algorithms that I've used in the past that would make it virtually impossible to retrieve the fingerprint from the card. This way, you would touch the card; it takes the fingerprint for a millisecond, transforms it and then compares it against an opposite fragment—like putting two pieces of a broken cup together. I am grossly oversimplifying the process for the sake of example but it suffices that I've seen it work. The issue is not whether it works but rather under what authority a merchant could demand a fingerprint?

Smart cards, in whatever form they may take, present a wide range of privacy and security issues that have yet to be fully resolved. Among the more prevalent concerns is the high volume of transactions that these cards generate are generating an exponentially large data trail on consumers. But that is only the beginning. As the cards become increasingly more intelligent, there is the risk that the cards may begin to harbor additional functionality that wasn't originally planned for or agreed to. There is nothing to indicate smart cards will not be equipped with transponders that would provide the ability to pinpoint a person's location along with their purchase history. Every credit or debit card transaction already possesses an audit trail that includes where the transaction occurred as well as other details and I can say, from professional experience, that banks do go to great lengths to make this data is very secure.

Another valid concern has been the gradual expectation that credit cards and driver licenses have become the de facto form of identification required to complete specific classes of transactions even when there are no charges incurred against the card—just try renting a hotel room or a car without a credit card and will appreciate the gravity of this fact. Conversely, it is only fair to point out that merchants, particularly those noted, need a form of collateral or identification in order to entrust a consumer with a very expensive asset. Unfortunately, this raises the risk of fraud, theft, and abuse as well as many of the privacy issues that I've already raised. It also infers that the subset of the population that does not have a credit card will be denied said services. In a few cases, several retail chains including video stores demand a driver's license—not to verify your identity but to scan the license for account information. The question should be asked; under what authority do a retail chains have to demand a driver license?

Cash, in the eyes of many government agencies and corporate interests, creates a lot of problems that are rectified by the proliferation of smart cards in exchange for the conveniences offered to consumers. So why is cash such a bad thing? It denies the ability of governments (or whomever) with information that is considered essential to the tracking of transactions. To the extent that smart cards and debit cards replace cash transactions, there is no diminution in the ability of law enforcement agencies to trace the flow of funds, and to use payment systems as locator mechanisms for persons under investigation. Authorities in most countries already, by law, track and report fund transfers over $10,000 in the United States, Canada, United Kingdom, and Australia, as well as other countries. As the cards have higher limits, support multiple currencies, and allow person-to-person transfer of money, the level of concern will rise. Totally anonymous transaction vis-à-vis cash, under these circumstances, are not favored because they do not provide a data trail. This isn't just abort narcotics smuggling or terrorism; cash is also the conduit of the underground economy that undermines the ability of governments to raise revenues. Likewise cash undermines the ability of marketers the rich fountain of information afforded by smart cards.

In spite of the privacy concerns there is nothing to suggest that the public should suddenly cut the millions of cards in circulation or be overly paranoid—neither will resolve the privacy issues with smart cards. But the use of cards comes with it an unseen cost to your privacy. Though each transaction you use with your card may appear to occur between the clerk and yourself, there is no telling what will happen to that data. Given the public's voracious appetite for credit and conveniences coupled with the trend away from cash, the only way to provide the conveniences of the card while providing sufficient security and privacy to cardholders is to ensure banks and other providers adhere to the strictest standards of confidentiality and encryption.

4.4 RFID: The New Generation of Radio Frequency Identification Chips

"Technology is a way of organizing the universe so that man doesn't have to experience it."

—Max Frisch

There is yet another battle beginning to form that has yet to make its way into the mainstream press that involves tiny chips capable of transmitting information about the item they are attached to. These chips, referred to as Radio Frequency Identification or RFID chips, can be as small as a grain of rice but include storage capacity and an antennae and can be used in everything from security passes to inventory control. But with these advances, privacy concerns are also beginning to surface that raise the specter of RFID chips being used to track people rather than things.

The danger of RFID stems in part from their exceptionally small size and the seemingly invisible nature of their operations. Firstly, it would be difficult for the average consumer to even known they are carrying a RFID in their clothing or shopping bags, and secondly since the chips are communicating to receivers over radio waves, consumers would be none the wiser. As is often case, RFID chips are not evil and there are many valuable applications but the question is one of self-control. Will the marketplace be able to constrain itself from allowing the technology to invade the privacy of consumers in a manner that leaves consumers with no opportunity to consent to their use.

Before getting into the privacy and policy issues surrounding radio frequency technologies, you need to understand what they are and how they work. Radio Frequency Identification is a form of wireless technology with the potential to offer *contact-less* services such as checkouts, payment systems, security systems, and automated data collection.

These tags have the ability to work in different modes including read or write and passive or active sites and do not require line of sight with the reader that are usually mounted on the walls or pillars. The tags can store information without a battery and can come in virtually any size or shape. Inside every RFID a microchip is attached to wireless transducer and that antennae essentially can read or write to readers in its proximity; in fact, the readers generate a radio frequency field around their antennae that in turn provides power and time functions to the chips in its area.

The chips are not yet capable of storing a large amount of data and currently hold approximately 16 kilobits of information; usually enough for identification names or inventory numbers and locations—some codified number schema. The chips themselves are quite useless if there is no means of interrogating the data and RFID readers fulfill this role. Once the data is acquired from the chip, the RFID reader can pass that data to all sorts of computing facilities for operations and analytics. In the case of passive chips without their own power sources, readers emit an energy field that "wakes up" a passive chip. Active chips are more sophisticated and have their own power source and transmit periodically in the Ultra High Frequency range of 900 MhZ to 1.9 GhZ.

So far the cost of RFID chips remain very prohibitive for all except the largest retailers but keep in mind many, like Walmart, have ordered their primary suppliers to become RFID enabled and compliant and it is fair to assume that all of their suppliers will have to meet this requirement to continue doing business with Walmart. If the applications remain limited to logistics and inventory then there is no problem and we would expect any company to leverage on best-of-breed technologies to maximize operational efficiencies. It is when the technologies reach the store floors that the problems begin; remember that a RFID is small, very small—as in less than a cubic millimeter, which means it can be easily stitched into clothing or attached to boxes.

RFID technologies have become indispensable in the identification and location of pets and it should make you wonder when and whether it will be used on the location of people. If it is still a little murky, consider a future in which every object that we use or consume or own will be embedded with RFID and they in turn will be connected to the Net for collection and analysis. In fact, you could already be "exposed" to the technology without even knowing it. In Canada alone,

- *If you use the 407 Highway in Ontario then you've already been exposed to RFID chips that are built into the transponders.*

- *If you buy your gas at Esso and use their SpeedPass Program then you are using RFID where a transponder in the pump recognizes the identification signal sent from the key tag you were given.*

Considering how easy it is to implant RFID chips, guidelines and controls are required to make sure the market is playing fairly and there are differences of opinion as far as how this should be implemented. Many in the privacy field including the ACLU, the Center for Privacy and Democracy, and Consumers Against Supermarket Privacy Invasion And Numbering (CASPIAN) have all testified before a variety of government subcommittees and pushed for broader privacy laws that would enshrine a series of general protection principles that I have touched on in different chapters. Walmart, Proctor & Gamble, and many in the RFID industry have been just as quick to the floor and have countered, not surprisingly, that market place should set its own rules and allow self-regulation to govern their behavior. But I have already made my opinion known with respect to the efficacy of self-regulatory regimes—they are ineffective and fail to address the conflict of interest when an industry has attempted to police itself.

There is mounting evidence that RFID's will become pervasive everywhere we go and it is, as a matter of record, the largest companies that are driving this push to use this technology. With Walmart pushing suppliers towards compliancy and

Gillette ordering over 500 million tags, as well as indications that governmentsare beginning to see the value in these chips, this is a technology worth watching.

Many of the applications being considered for RFID have absolutely no bearing whatsoever on privacy and tend to focus more on leveraging existing CRM database technologies, point-of-sale systems, warehousing and distribution, inventory control and security, supply chain management, ERP systems, and a large number of enterprise-level systems. The development of new vertical integration markets for the technology is further noted by the number of "big" players that have gotten into RFIDS such as NCR, Texas Instruments, and of course IBM. It is still very early in the game and the possibilities are only starting to make their way from the laboratories to the boardrooms so it remains to be seen how well the technology is adopted; but if market indicators provide any barometer of activity, RFID is going to be huge. Put in financial terms, if the RFID industry in its infancy saw shipments of $1.2 billion in 2002 according to Business Solutions Magazine, and estimates from the Yankee Group place the market value around $4.2 billion by 2008. There are varying estimates of how big this market will be and the technology is not without its technical and public relations hurdles.

RFID is not a perfect technology and there are still technical hurdles that must be solved. So far the greatest challenge has been the cost of chips—not a major issue for a small animal hospital dealing with a few hundred implants, but a very major concern to companies with millions of items to be tagged. There are also several standards including Class Zero and Class One; the standards are changing so fast it can be quite risky while waiting for the standards to stabilize. Even the placing of tags can be tricky because the signals do not penetrate well through liquids or metals. Lastly, the storage capacity on the chips is very small at the moment but that is likely to change. For the moment the best thing to store on these chips are some for of identification numbers such as lot numbers, part numbers, and similar identifiers. It is almost impossible to predict how RFIDs will evolve or how the technology will be applied—it can only be assumed that the chips will become smarter and cheaper, the storage capacities larger, and so on.

In terms of guidelines for development and usage, they are largely the same as they would be for any potentially privacy-invasive technology. Internally, any organization that is considering RFID technologies should be conducting privacy assessments long before the first line of code is written; what is the purpose of the new application? Are there privacy issues? What is the scope of data being collection, its retention, usage, and disposal going to be? All of these questions need to be answered during the design phase before the chips figuratively and literally hit the floor.

Secondly, but no less important, is how this information will be convey to customers or employees. If RFID will be deployed to provide building security access, what are the limitations of that access? Will the chips track every movement of an employee within a facility through the use of multiple readers at various points in the building? Or will it only be used to ensure employees are authenticated when they enter or leave a facility. If the technology is for use in retail environments, policies must be clearly and publicly posted that warn customers that RFID technologies are in use so that customers may in turn make their own decisions.

Taking a more skeptical view, will consumers really understand what they are reading even if it was posted properly? Most of us have trouble programming the VCR even though it's been around for decades, what is the hope that the average consumer is going to have the ability to understand the implications of an advanced technology let alone possess the requisite interest. That isn't to demean the predominately non-technical population—including many of my readers—it is, instead, to emphasize the point that the majority of the population is *not* engineers and there must be an effort to educate the public in clear, unambiguous terms.

There are certainly other steps that can be taken to address the privacy issues associated with RFID. Organizations looking implement tags must be as forthcoming and open with their respective customers or employees as possible—you can be sure that any attempt to covertly use tags in any environment will eventually be discovered no matter how clever the scheme. The same operational and personnel controls that apply to sensitive information systems must extend equally to tag technologies—in many cases those tags are identifying people and that raises many legal and privacy issues that require a higher degree of security and oversight than is normally the case.

In terms of what is stored on a tag and its relationship with a customer or employee, every effort should be taken to break the link between the item and the person. It is one thing to track the movement of an item and a whole different situation when it involves tracking a person. The public, despite the previously noted lack of technical expertise, will nonetheless fully understand the implications of being tracked and will not be returning to the store if there is sufficient media exposure. If the tag absolutely requires customer information written to the chip, every effort must be taken to either encrypt or block rogue readers.

Any new technology that has the potential to violate a person's privacy will be met with a healthy dose of skepticism until the objections are resolved or the market can prove itself capable of self-regulation. Historically, however, the market has rarely shown itself capable of resisting the temptation to resist privacy and for that reason legal and statutory protections will be required to reign in the growth of RFID applications. RFID is really a fascinating technology and it has many legitimate applications like inventory control and loss prevention. It is only when "tagging" is covertly or forcibly pushed on the public that problems begin to surface and it would be considerably easier if RFID vendors and their clients tackle the issue before it is tackled for them.

4.5 Location & Tracking

"The real danger is not that computers will begin to think like men, but that men will begin to think like computers."

—Sydney J. Harris

As if it wasn't enough that there are so many means of determining *who* a person is and *what* they purchase, there is a lot of research into how to determine *where* a person is at any time of day ranging from location systems to capture license plates on the 407 in Ontario to cellular and GPS tracking. It may not seem like a big deal but it is worth thinking about a days worth of movement data reveals about a person's behavior. By combining the GPS coordinates of a person's movements combined with other data sources like Geographic Information Systems, it becomes quite easy to discover virtually any aspect of a person's dialing routine.

Assuming an organization wanted to track the location of people or where they shop, or where they go to school, the technology exists if they were so inclined. And so far I am referring primarily to the GPS tracking of vehicles—I not yet entertained a world of implanted chips. Far from some strange science fiction, the tagging that once began with pets is now moving to Alzheimer's patients. This is not some mandatory requirement of the state, this is a private sector service—and there are several thousand senior citizens on a waiting list waiting for this implant.

Perhaps you need an example a little closer to home. Unbeknownst to the majority of cell phone users, there is always the potential that phone companies and governments can determine your location through your cell phone, even when it isn't being used and that is prompting civil libertarians and privacy advocated to warn the public of the dangers associated with location technologies. Of course the issue is not that simple—you would want police to be able to locate your cell phone if you were in distress or calling 911 but you *wouldn't* want anyone to determine your whereabouts unless it was within reason. As GPS is incorporated into more cellular phones, it is likely we will witness increased surveillance of our cell phone calls under the umbrella of national security or some other convenient logic. It isn't the telecommunications giants that are at the heart of this move; it is the government that is pushing to incorporate location capabilities into cellular devices and if you bought your phone after December 31st 2002, then it is already equipped for it.

The original rationale for the new FCC requirements was based, presumably, on the concern that traditional cell phones were not as reliable as land line phones. However, it still raises the specter of unfettered tracking unless it is either countered with consumer options for disabling the service, limiting its use, or regulating how and when a phone can be tracked. For all the talk of GPS and advanced technologies, phone and cellular carriers can already determine a person's position through a combination of signal strength and triangulation—the new devices simply make the process more visible and accurate.

A logical progression in this discussion is a quick look at GPS since GPS is becoming more affordable and easier to use. The Global Positioning Satellite network consists of a constellation of 27 earth-orbit satellites that was originally developed by the U.S. Military who continue to operate this network that now includes a large number of civilian applications. Each of the solar powered satellites weighs approximately 4000 pounds and circles the earth twice a day with the orbits designed so that there will be at least four visible satellites no matter where you are on the planet. A GPS receiver, the device carried by a person or vehicle, locates four of these satellites and then determines its distance to each of them;

once the distances have been acquired the receiver performs trilateration equations to determine its exact position.

Fundamentally, pinpointing a point in three dimensions isn't that difficult but it's a little trickier to visualize. Imagine the radii from the examples in the last section going off in all directions. So instead of a series of circles, you get a series of spheres. If you know you are 10 miles from satellite "A" in the sky, you could be anywhere on the surface of a huge, imaginary sphere with a 10-mile radius. If you also know you are 15 miles from satellite "B", you can overlap the first sphere with another, larger sphere. The spheres intersect in a perfect circle. If you know the distance to a third satellite, you get a third sphere, which intersects with this circle at two points. The Earth itself can act as a fourth sphere—only one of the two possible points will actually be on the surface of the planet, so you can eliminate the one in space. Receivers generally look to four or more satellites, however, to improve accuracy and provide precise altitude information. In order to make this calculation, then, the GPS receiver has to know two things:

- *The location of at least three satellites above you*

- *The distance between you and each of those satellites*

The GPS receiver figures both of these things out by analyzing high-frequency low-power *radio signals* from the GPS satellites. Better units have multiple receivers, so they can pick up signals from several satellites simultaneously. Radio waves are electromagnetic energy, which means they travel at the speed of light (about 186,000 miles per second, 300,000 km per second in a vacuum). The receiver can figure out how far the signal has traveled by timing how long it took the signal to arrive. These in turn are used to determine your location, which can be converted into "user friendly" coordinates or visual mapping interfaces.

Regardless of whether I am referring to cellular or GPS, the real concern is the question that is conspicuously absent; who will have access to the data that locates where you are? What policies exist or will exist to protect privacy? What assurances are there that companies will not sell location-specific data to third parties? As a matter of historical record, carriers have not been blessed with a sterling record. If carriers turn over information in response to a legitimate court order then we must accept the integrity of the courts. And as more people embrace GPS without considering the consequences of being traced, the need to regulate access to this information will become crucial.

One telling application that won the surprising endorsement of the American Civil Liberties Union has been the use of GPS in the tracking and surveillance of paroled convicts. These electronic tracking systems essentially follow released parolees wherever they go and compare their movement patterns with past criminal convictions and new crimes committed wherever they are. The historical movements of a convict can then be used by law enforcement, crime analysts, and parole officers to determine if a person has possibly committed crimes or violated the terms of their parole. If a person had been previously convicted of a sexual offence and forbidden from going near schools or parks, the real time GPS tracking systems could prevent many offences from occurring and the same can be said of restraining orders. The knowledge that a person is being watched might also act as a deterrent.

Seldom mentioned is the possibility of dramatically reducing the prison population by releasing non-violent offenders and thus reducing the costs of incarceration and allowing these people the opportunity to reintegrate into their communities under carefully controlled conditions. This can be an enticing economic model provided the release of offenders is based entirely on their behavior and not on the reduction of costs. Still, having offenders pay $6 for the tracking service versus the estimated $60 per day is a win-win for the convict, the community, and the prison system.

A more compelling and sobering reason to allow the use of surveillance satellites is missing children. Based on the *National Incidence Studies of Missing, Abducted, Runaway, and Thrownaway Children (NISMART-2)*, released in October 2002, there were 797,500 children abducted in 1999. A majority of those were attributed to family abductions but there were also 58,200 "non-family" abductions that unfortunately are rarely solved. There are several systems already in operation that combine tamper-proof bracelets or other mobile tracking devices and centralized operations centers that have the potential to change the dynamics of abduction in favor of the parents and law enforcement. Theoretically, the location of a missing child could be completed before an abductor could get out of the parking lot. These systems will increasingly become visible in malls, amusement parks, and metropolitan areas as the technology continues to improve.

With the provision that location systems be governed and constrained in their application, these systems have the potential to save many lives and I wouldn't want to ban location technologies. If drivers are given the right to consent to GPS locators, many accident victims could be located considerably faster and that would save a lot of lives. Similarly I am equally excited about the development of "child find" systems so long as only the parent's control access to the child's locator beacon or the right to allow law enforcement access. Without a doubt, embedded GPS locators would make it much easier to locate missing or abducted children in a matter of minutes and not days. The common element to these systems is control and consent and so long as consumers control access to location information and consent to the collection of data then there are many positive aspects to location technologies. It is only when locators are used covertly and surreptitiously that security and privacy issues need to be resolved.

There are already "private investigators" offering services to track wayward spouses using covert GPS receivers—if you do, I'd suggest checking under your car next time. It would take less than two minutes to roll under your car, implant the device and no one would be the wiser. But GPS coordinates are not of much use by themselves—they're power is increased exponentially when they are coupled with mapping and GIS Services which can superimpose the GPS coordinates over maps and show what was at a particular location where a person parked for three hours.

Well, here we go again; another case of a technology that has the capacity for great good and yet also the capability for considerable harm; and for that, the same caution holds true—location services and technologies have the ability to improve and save lives but protections are needed to ensure the ability to locate a person is not abused. I am personally not concerned with applications that can locate pets, children, or an Alzheimer's patient—who could argue with that. What raises concerns is the very real possibility that individuals could be traced without their knowledge or consent.

4.6 Medical Privacy & Genetic Testing.

"If knowledge can create problems, it is not through ignorance that we can solve them."

—**Isaac Asimov (1920–1992)**

For one that follows science so closely, medicine has always been one of my favorite pursuits and the subject of medical privacy and genetic testing is especially germane to this book and an appropriate end to a chapter on advanced technologies. The protection of medical information is unquestionably the most important of all private domains and surely it is expected that this kind of information would be restricted to one's family and physicians—and you would certainly demand the same of diagnostic tests. But, again, if it was that black and white, there wouldn't be a segment dedicated to it.

To put this into perspective, try to imagine two people that are applying for the same job where one is very attractive and the other is not. If it is assumed that the two candidates are equivalent in education and experience, which do you think will be given preferential treatment? Barring the contemporary mantras of equality and the doctrine of political correctness, the more skeptical amongst us cannot be blamed for believing the more attractive candidate will have a better probability of success—legislation notwithstanding, its hard to believe the law can compete with human nature.

What if the decision to hire someone was based on something far more intimate than casual appearances? What if such decisions were based instead on a DNA sample or the possible predisposition toward conditions that the candidate didn't even know they were at risk of developing? These types of decisions run perilously close to the abyss of eugenics and define one the strongest arguments for the prevention of disclosures of medical facts. This is not the stuff of science fiction and as genetic testing becomes more sophisticated it will have to be confronted.

Before I get ahead of myself, any discussion of medical privacy or genetic testing logically necessitates a quick (and crude) review of basic biology. To paraphrase an entire textbook, every cell, with the exception of red blood cells, contains genetic information that defines who we are. That genetic information is composed of 50,000-100,000 genes that are contained in deoxyribonucleic acid

(DNA) that takes the well-known double-helix structure (actually it's been proposed that there is three strands not two).

The genes in turn are comprised of over three billion chemical base pairs defined as (G) Gaunine, (A) Adenine, (T) Thymine, and (C) Cytosine that combine into defined pairs like A-T and G-C. A gene is further defined as a series of base pairs located at specific locations of the chromosome. From this, every human cell has 23 chromosomes from the mother and 23 chromosomes from the father that combine through reproduction to form the child's chromosome. The last one is the autosomal or sex chromosome that determines your sex.

Genetics and biology play a key role in diseases and illnesses with over 5,000 genetic disorders that affect, on average, five percent of the population. Broken down a little further, genetic diseases can be classified based on their cause: monogenic diseases that are caused by a problem in a single gene (Cystic Fibrosis), polygenic diseases that are caused by problems in multiple genes (Diabetes Mellitus), chromosomal aberrations that are caused by structural errors in chromosomes (Downs Syndrome), and finally non hereditary diseases like cancer that are caused by mutations in the cell.

To discover the existence of these conditions, medical science thus far relies on three predominant techniques that make use of probes and markers including genetic testing, genetic monitoring, and forensic analysis. Genetic probes look for a specific gene that causes a particular condition and where the base sequence is already known. Genetic markers are used to determine the probability that a condition might exist. Since these genes are often in proximity to a target disease gene, their existence or presence indicates the possibility of disease and are very useful when the base sequence is not known. It is important to emphasize that genetic markers in particular only signify a probability that a genetic condition *may* exist. That is an important fact when arbitrary decisions are being made with respect to the well being of an individual.

The science is important but so is the undercurrent of curiosity. Think back to the last time a friend or family member was ill and the first inclination is to ask how the person is doing followed immediately by what it is that is causing their illness (or vice versa). To ask the former is natural but to ask the latter crosses the boundary into a very private domain. With families or friends, this is understood to be a genuine concern for one's health but the same is not extended when these questions are asked beyond this inner circle and the motivations are not so benign. The institutional curiosity within an organization is compounded exponentially and acts a force multiplier except the reasoning is quite different. Organizations are not biological entities and their motivations are not driven by compassion—the drive for information is to either gain competitive advantage or reduce risks. Normally this is an external pursuit aimed at an organizations rivals or customers but this process can take an internal twist when the energies are directed towards employees or potential customers.

Imagine the possibility of insurance companies denying you coverage because of a slight probability that you *may* eventually develop Parkinson's Disease, that a bank might turn your down for a mortgage because it detected a chromosomal defect, or that an employer determined you have a hereditary probability of cancer. Now take that a step further, imagine that anyone could have access to your DNA to determine whether you get a job, go to school, get social benefits, or even housing. The science of genetic testing and monitoring coupled with the mapping of the Human Genome will mark the most invasive assault on human dignity and privacy in history. Do I have your attention yet? The mapping of the Human Genome isn't to blame; it will be the gradual push towards national databases of DNA that marks that point. Genetic engineering, testing, and monitoring has the potential to save millions of lives as it in turn leads to the discovery and development of new drugs and treatments for a great many diseases. But it must be controlled to prevent organizations, no matter their intent, from leveraging on science to achieve their own self-interests.

For those readers not familiar with the biological sciences, it can be difficult to come to terms with the prospects of genetic testing but this is not a figment of the imagination and in many ways, as pressure increases to apply these advances to social issues, it will become an extension of drug testing that is already rampant in the United States. There are legitimate arguments for mandatory drug testing of specific occupations such as police officers or pilots for the obvious reason that an addiction to narcotics endangers the individual and the public. The same could be argued of any profession in which there is an element of public trust.

Narcotics are a horrible blight upon society but the "War On Drugs" reminds me of Nietzsche's warning that he, who fights a monsters, becomes a monster and this is reinforced by the increasing number of high schools and private companies that have attempted to enforce mandatory drug testing. How long will it be before it is deemed mandatory for the population? How long before drug testing mutates into genetic testing? When the social interests of government or the economic interests of the private sector either influence the direction of medical science or, conversely, base decisions on medical conditions, the consequences extend not only to privacy but also to the right to survival including the right to housing, food, and employment. That any organization could deny a fundamental necessities to a seemingly healthy person based on the mere possibility of a defective gene speaks volumes of the risks associated with genetic testing. So far statistical data indicates that organizations are reluctant to pursue genetic testing due to the negative backlash that would accompany such a movement.

It is from this quite clear that genetic engineering in general and genetic testing in particular will have enormous benefits to society not the least of which will be the eventual eradication of genetic diseases that have plagued mankind for thousands of years. But everything has a price tag, and the price of progress is the possibility that the science can be harnessed for less than humanitarian purposes.

4.7 The Future of Technology & Privacy

"I don't think the human race will survive the next thousand years unless we spread into space."

—Stephen Hawking

There is a demarcation line between the technologies that we are commonly exposed to and the science fiction of the near future. With the exception of data mining and smart cards, the technologies that I have brought to light may never have a direct impact on you as an individual but it will eventually have a collective impact on the population. Then again there is nothing to say when you will be suddenly asked for a fingerprint or blood sample. But most if not everyone has at least one credit card in their wallet and there is little doubt your name sits in many databases so you are affected by technology whether you realize it or not.

As for location services and genetic testing, these are a little further on the horizon and I doubt you will see them for some time; newer vehicles after 2005 will all have the ability to be tracked in the United States so your next car will in all probability possess the potential to be located via GPS. The infancy of genetic testing coupled with the broad opposition to its use and the constraints of the law will slow the pace of genetic testing and monitoring but it will not completely halt its development. But many firms already have mandatory drug testing programs and it wouldn't be much of a stretch to impose genetic testing.

It is virtually impossible to say how technology will change and to put it nicely; the human species does not have a good track record when it comes to predicting the future. It is often quite amusing to look back with the gift of hindsight at the predictions that have been made by some of the brightest minds in the world. And in most cases, to speculate that something is impossible more often than not becomes common place. During the 1920's, the following was once said...

"The wireless music box has no imaginable commercial value. Who would pay for a message sent to nobody in particular?"
—Associates of David Sarnoff, manager of an early US radio network, 1920s.

Though the explosion of mobile devices makes such statements appear unimaginative, perhaps all predictions should be qualified as a function of the time in which they were offered. At that particular time in history, when the technology was very much in its infancy, it must have been difficult to envision what are now considered conveniences. If you really want to see unimaginable, exponential growth you need only look at cell phones with particular attention paid to the rate of change over time. Based on the number of cellular phones per 1000 persons per country, these statistics give you a clear picture of how fast technology can proliferate.

Country	1985	1990	1995	1997	2001
United States	1.4	21.1	128.4	206.5	450.8
Algeria	0.0	0.0	0.2	0.6	3.2
United Kingdom	0.9	19.4	97.9	149.8	770.4
Hong Kong	0.8	24.4	129.7	339.7	859

Never mind the numbers for the United States, look at the growth of cellular usage in the United Kingdom and Hong Kong. From 1997 to 2001, a mere four years, the United States doubled in the number of phones, which was completely dwarfed by the five-fold increase in the United Kingdom. Both countries, contrary to how wired you *think* Americans may be, experienced rates of growth that far exceeded North America. And the track record for predicting computing growth is not much better. As Thomas Watson had said in 1943, " *"I think there is a world market for as many as five computers"*.

To think forward fifty years later, it is somewhat ironic that science and society are confronted with a problem that was unthinkable in 1943. Not only must society establish a code of conduct for people, we must now establish a similar rule set for *machines*. Intelligent, autonomous, and mobile robotics and artificial life forms are going to raise some very serious questions in a relatively short period time.

Just as a teaser, the Utah-Brain is a fully autonomous (independent) robotic cat under development at the University of Utah that claims to have 75 million artificial neurons (double the number of the entire Republican party) and that number is anticipated to grow to over a billion "cells" through an adaptive self-evolving neural network (artificial brain). If the theory holds and I suspect it might, this synthetic hairball will have the ability to see, hear, touch, feel, think, and ultimately make decisions for *itself*. Other robots like ASIMO (Advanced Step In Innovative Mobility) claim the throne as the most humanoid robot ever developed (so far) including the ability to walk and climb stairs, and yes, dance—not too bad for a bucket of bolts.

I watch on occasion with no small measure of bemusement as so many people obsess about alien life on other planets when in fact "alien" life forms, in form of the aforementioned cat, are being created in laboratories as we speak. And while it makes for fantastic headlines, if they should ever reach a state of self-awareness, what more constructions have been layered into them such as privacy and respect for human beings etc. Even if they never achieve this vaunted state of consciousness, they will likely be able to mimic human cognitive processes—maybe at some point they might even have the ability to say "No". Quite true, we are not at that point, but how much money would you bet against it?

Humans, too, if there is a gentle way to put this, are in the process of being re-engineered and the expression "Are you wired?" is going to take on a brand new meaning. To give you some idea of what this really means, ask yourself how many chromosomes you have. The answer is, or more to the point, *were* 46. At least until Chromos Molecular Systems developed a 47th artificial chromosome that has been introduced into mice. Perhaps in 50 years, we will need as many upgrades as our PC's.

There are thousands of researchers experimenting on areas of science and technology that will force society to completely rethink many of the most cherished assumptions about quality of life, freedom, even the meaning of life itself. To name just a few, one project involves integration of the human nervous system to robotics that will allow patients to regain feeling and senses, another includes a blind patient that had his eyes *replaced* with a digital video camera where electrical signals are processed by a microprocessor that transmits the signals to the nerves in the visual cortex and has given the patient the gift of sight—it is still archaic and rudimentary system that looks more like stadium lights to the patient but it is just the beginning.

I am sure there will be some readers that find the mere thought of such things absolutely abhorrent and unethical; after all, these people are not laboratory animals. True, but to that blind patient with no hope of sight, this presents a no-lose opportunity to regain even a rudimentary sense of sight that we take for granted. Rest assured he would not be losing any sleep over the ethical questions that arise from this research.

Though I am reminded of the folly of those that have made predictions, it is still tempting to consider what the future might bring. And besides, if it weren't for our creativity, imagination, and a willingness to take risks, we would, as I pointed out earlier, still be figuring out how to make fire from flint. How much of this is possible or science fiction is an open question but I would be more inclined to put money on the possible than the impossible. Granted, contemporary medical science can even solve the cold let alone the idea of artificially intelligent immune systems. I am also quick to caution that there are those that have and will always make it their mission to prevent the advance of science and the litmus test for science must rest on objective science, not on religious or ideological dogma and intolerance. Society still requires an ethical framework to make sure these advances are not abused and within the body of that framework, the questions of privacy, dignity, and autonomy must be addressed. The point of this whole discussion is to make it abundantly clear that what the future holds is anyone's guess and, if some of these examples are any indication, it is also clear it is more a matter of when than if.

As I continued to edit this manuscript (yet again) in the Spring of 2005, a flurry of developments in embryonic stem cell research began to emerge that has ignited that firestorm yet again; this time teams in the United Kingdom and Korea were able to successfully clone a cell through somatic nuclear transfer and even personalize the stem cell for a specific patient—the point being, things we once thought impossible are here and we are indeed playing God. That these developments happened in other countries ought to serve as a reminder that these sciences will continue to develop whether we are onboard on not. Rather than isolate ourselves or pretend it will never happen, it is far more prudent to accept progress and define the framework for working with it.

5

Biometrics

5.1 Introduction to Biometrics

"Biometric technologies have substantial implications for privacy and so are more than a passing interest for privacy advocates"

—Susan Hart, Financial Economist,
Office of Critical Infrastructure Protection and
Compliance Policy
U.S. Department of the Treasury

Within the last four years the topic of biometrics has garnered its share of media attention in light of the war on terrorism, a reinvigorated focus on domestic security, and a rethinking of organizational security. But the discernible lack of dialogue with respect to the social impact of biometrics, including the erosion of privacy, warrants a focus on this technology. Technically any of the advanced technologies that were outlined in Chapter Four would've merited the equal coverage but my own experience in this field and its relevance to privacy tilted the balance in favor of biometrics. The use of fingerprint scanners and retinal readers has so far managed to slip under the radar of public scrutiny and the increased usage in public applications deserves additional attention.

For a majority of readers, the best way to explain biometrics is through the invocation of populist themes that are often found in movies and have even made their way onto the occasional cable news segment. Unfortunately the coverage or use of biometrics on camera tends to glamorize and oversimplify the science but some coverage is better than none. With that in mind, many movies have incorporated biometric readers to convey futuristic vision of gadgets and the best examples of biometrics in action have been the scenes from the Mission Impossible and James Bond series.

Rather than condemning Hollywood for taking liberties with the science, I am actually quite pleased—it is often the only place that the general public is introduced to new technologies, and as I said, the exposure is healthy. Still, there is the risk that these fictional portrayals may lead readers to think this is the stuff of science fiction when in fact palm readers and retina scanners are beginning to proliferate in many corporate and government institutions. For many years such advanced means of authentication were the exclusive domain of the military or spy agencies but they have seeped into the public and private sector—even if you don't know they are there.

There are, regrettably, a number of problems with biometrics not the least of which are the violations of personal space and the human body that have been cast aside in the pursuit of science and security; for one that wrote his doctoral work in this field I can personally attest to the frustration of explaining the technology let alone the privacy issues which is quite often met with indifference, confusion, and apathy. It is also incumbent upon me to profess that I am not an innocent bystander in this drama but an active, albeit small, participant; though not deployed in the field, my work would have, if not revised, contributed to the erosion of privacy—alas, though barred from discussing my work, privacy became a central focus of that research and I may rest comfortably in the knowledge that my contribution to biometrics was not absent of conscience.

There have been calls in some quarters for a moratorium on biometrics but, as much as I respect highly the opinion of the leading minds that have been arguing for a halt to this proliferation, I do not share a belief in the need for an absolute freeze but instead a fair assessment of the social impact of new applications before they are unleashed into the public domain. Excluding a litany of abusive applications there are many positive uses for biometrics and it would be a shame to destroy the good with the bad—I'd like to think we can exercise a degree of rational selectivity to weed out those that are dangerous while preserving those that preserve privacy and freedom and make a valid improvement in security

Society must first come to terms with the realization that the convergences of the key technologies I've discussed so far bring society precariously close to a surveillance state and many have argued that we are already there. Roger Clarke, one of the most notable figures in privacy, had suggested there were three conditions that must be met that would give rise to a surveillance state. Briefly stated, they are as follows:

- *The first condition was the creation and development of data management systems that could record personal information on a large scale. This has been achieved through a variety of Very Large Database (VLDB) architectures since the 1980's and into the 1990's with the emergence of data warehousing and new data mining algorithms.*

- *The second condition was the emergence of an overarching networking infrastructure or backbone that would connect these isolated data repositories—the Internet has taken care of condition two.*

• *Finally, the means of uniquely identifying every person on the planet has been achieved through the development of biometrics. Without getting into the data structures, rest assured there is ample storage and network capacity to maintain a record of every fingerprint on the planet.*

Things have changed dramatically in the past 40 years and what were once theoretical musings have given way to reality and the three conditions postulated by Clarke are in place—notably the *technical* means exist to store and track every human being on the planet. Databases that once held thousands of records now have the capacity to hold *trillions* of records and network capacity once measured in thousands of bits per second is now measured in the billions. For that matter, the average home PC has more power and storage than many of the mainframes of the 1960's and the network speed and storage capacity of large institutions are exponentially greater. The last piece of the puzzle was the third condition; the existence of the unique identifier, and the emergence of automated biometric systems has fulfilled the third condition. The odds of a surveillance state are no longer a question of ability; they are a question of social change and political will.

The utmost challenge with biometrics is not so much that it collects information *about* us; it is that biometrics is collecting information that *is* us. It is one thing to ask a man for his name for even the law holds that a name is generally accepted as a matter of public record, but to demand that a person surrender themselves to such violations is very offensive. To provide any kind of biometric measurement requires the physical manipulation of an autonomous individual and it still a violation even if the process does not physically come into contact with skin.

Ultimately, the unchallenged evolution of this technology stands to rob the individual of their dignity and the right to remain anonymous beyond the normal transactions requiring identification. And if left unchecked society runs the risk of a day where the mere act of a drive to the store or entering your home, requires fingerprints. It is still quite farfetched but to deny the possibility is to deny the existence of the cameras that are already on the highway and at the intersection—the surveillance is slowly closing in on the individual home where a man's castle might well become his prison. This must be the penultimate dream of the paranoid that believe the populace is up to no good and the insecure amongst us that feel the need to control others. Despite the appearances that all of this is the direct result of scientific advances, the desire to uniquely identify a person is nothing new—the technology we have now simply makes this task incredibly accessible.

At a slightly more formal level, biometrics provides the ability to identify or authenticate an individual based on a multitude of physiological and behavioral characteristics using a broad range of technologies that are dependent upon a large number of factors including the intended application and the environment it will operate in. In terms of the attributes that are currently being used, though not exhaustive, the following provides an idea as to what can be used to identify a human being:

1. Physiological Attributes

    a. Standard measurements including height, weight, eye color, hair color

    b. Fingerprints, footprints, retina, Iris, skull geometry, palm geometry

    c. Thermal imaging, capillary patterns, DNA matching, facial recognition

2. Behavioral Attributes

    a. Voice, movement, keystroke dynamics, hand written signature analysis

    b. Other forms of movement including how you walk (gait analysis)

These are the two key categories that are regularly used to differentiate between the classes of measurement, although physiological attributes are far more accurate and common. Of the measurements in each category, fingerprints, retina scanning, and palm prints are obviously more prevalent than skull geometry or thermal imaging, and voice identification, which remains the most common of the behavioral attributes. There is a lot of research being conducted on behavioral attributes but physical markers remain the more reliable. However, measurements serve no purpose unless there is a means of using them and when it comes to biometrics the objective is to either identify or authenticate a person.

First of all, there is a distinction between identification and authentication. Biometrics is, not surprisingly, about security; and the objective, regardless of the complexity, is to determine who you are. Authentication is the simpler of the two processes and seeks to establish a one-to-one relationship between a person claiming an identity against a known and stored identity to determine if the proposition is true. If the proposition is true then the claimant will possess attributes such as a fingerprint or palm print that was previously acquired and stored.

Though there is no limit to what this may be used for, the general premise to provide access of some kind whether it be for data or benefits or facilities.

Conversely, identification seeks to establish a 1:n search of an unknown identity (E.g. A face image captured in a store) against a large number of candidate images. In the latter case, a shoplifter is caught on camera but their identity remains unknown. To determine the identity through conventional means may require a person manually search through thousands of "mug shots" to see if there is match. The proponents of biometrics claim that technology can solve the problem by automating this search—rather than a person, a computer can search through every image and locate similar or identical features and provide a list of possible suspects. Interestingly, many systems only provide candidates—it still requires a human decision to determine if they have the right person. And that is how it should be.

Biometrics may come across as a strange fiction but the technology is gradually making it into workplaces, government services, and it needs to be brought into the light of public scrutiny. Biometrics is also about so much more than fingerprints and ridge bifurcations—it is about the invasive search of the human body and for that reason the privacy implications of this technology is staggering. To capture measurements of the human body constitutes a fundamental violation of civil liberties and defines the boundary between acceptable and unacceptable applications of science. There are a lot of questions that need to be answered before we allow anyone to ask the public to blindly submit to fingerprint scans.

5.2 Types of Biometric Measurement

"Hegel was right when he said that we learn from history that man can never learn anything from history."

—George Bernard Shaw (1856–1950)

The need to authenticate and identify a person is by no means new and support of our efforts to identify people can be found throughout history. There were no computers in the 17th century let alone the 2nd century so there had to be other means of making sure a person was who he or she claimed to be. And while it is barbaric by contemporary standards, the most common methods included tattoos, brandings, scarring, and even amputations—not exactly the most enlightened periods in human history but it was crudely effective. Other evidence has included many imprimaturs such as fingerprints discovered in clay tablets during the Tang Dynasty that indicate more progressive means of collecting human measurements. Time has moved on and there is, as of this point in our development, virtually no body part that hasn't been scanned, acquired, and tested—some with more success than others. This segment examines the more accepted types of biometric measurements that you are likely to come across in the news so that you have a better idea how each one works. And what better place to start then fingerprints.

FINGERPRINTS

Most of what the public understands of fingerprints has come from television police dramas like "CSI" that has come to reshape the public's perception of science and forensics. My favorite shows aside, fingerprints are the oldest, mature, and trusted of all biometric measurements and they have a rich history that has contributed to our knowledge. Fingers offer several characteristics that make them ideal for identification. They are very easy to scan, they are permanent, and they are unique. This uniqueness is formed at birth during the initial chaotic conditions of the embryonic mesoderm from which the patterns are developed—even identical twins have unique patterns though there are a few contrarians that have argued otherwise.

Although fingerprints look random at first glance, there are features and landmarks in the whorls and loops and it is these features that are used for identification. Though these features have provided sufficient uniqueness in the past, there are other features referred to as minutiae that are much more accurate and have become the de facto standard in "live" scan systems (AFIS). The minutiae are nothing more than the ridge endings and bifurcations but they hold considerable information. By combining ridge points, their location, and a system of orientation, even automated systems can differentiate between twins.

Fingerprint scanners are becoming increasingly common, even beginning to show up in stores, and it is more likely you will come into contact with this technology than you would the more advanced types such as retina scanners that I shall discuss later. It is relatively cheap, it is easy to setup, and it is trusted by the courts even though DNA matching is obviously more telling. Biometric fingerprint scanners are becoming so prevalent that they are now being installed into laptops and notebooks and portable fingerprint scan adapters are also available. In some ways this is a positive change and will ensure a laptop can only be used by one person. It is when the technology is coercively forced on consumers or employees that the problems begin.

HAND GEOMETRY

Hands have become a more common means of authenticating a person and it has gained in popularity as it has gradually matured and degree of accuracy has permitted sufficient trust to use palm prints in airports (INSPASS). Similar to fingerprints, palm prints are very easy to capture and have sufficient information to allow identification based on three-dimensional characteristics. This particular type falls somewhere in the middle of the technology spectrum and provides sufficiently strong performance to be used in airports and many other secure environments. Hand Geometry is not considered quite as robust as retina scanning but it is better suited to large populations of people with little or no experience with biometrics. It is also known to be less intimidating than fingerprints, which still carry the stigma of being used in law enforcement. However, it is not without some challenges—palms and hands are more easily scarred and damaged and any type of malformation or disfiguration of the hand would result in an error.

As is the case with all biometric devices, hand geometry involves an enrollment and verification phase. During the enrollment a user places his/her hand on sensor which is stabilized using "finger pegs" to ensure alignment. The image of the hand is captured several times and transformed into a template that is used for future verification. During the verification phase, the same user places his/her hand on a sensor for a "live scan" and that is captured to a temporary template which is compared against the stored template using a scoring engine to make sure you are the same person. The scanner, depending on the manufacturer, is a 2000 pixel charge coupled device that is capable of taking 90 separate measurements of a hand and then stores the measurements in an astoundingly small 9,000 byte template record. That small size is critical for the storage of millions of prints which we will likely see in the next few years as passports and other forms of identification are converted to biometric documents. The idea of a national identity scheme, while offensive to many, has been hotly debated and is beginning to take root in some countries. Personally, I suspect we will see a U.S. national identity scheme within 10-15 years and the erosion of privacy and civil liberties, coupled with the war on terrorism, will provide the push that is needed to bring this to fruition.

VOICE RECOGNITION

The human species have a remarkable, uncanny ability to recognize people simply by their voices, even when the voice is garbled or the speaker is ill, stressed, or have had their vocal cords damaged. It is a unique gift held by many mammalian species, which computers are only beginning to master to the point of providing a level of accuracy organizations would be comfortable using for the purpose of determining identity.

Using the voice as an identifier makes a lot of sense considering how easy it is to capture and transform into digital signals for comparison; it is also much less intrusive than fingerprinting and is well suited to integration with existing networking and telephony infrastructures. Perhaps most compelling, it is one of the most affordable types of biometric authentication available. Voice recognition uses a series of components including filtering and matching algorithms that capture and analyze the distinctive features of the voice such as the structure of the larynx, pitch, tone, cadence, and the spread spectrum frequencies of the sounds we make.

In the simplest of cases, a person can utter pre-determined words or phrases into a microphone and the system abstracts the measurements of the voice that are specific to those words and generates a template. When the users are to be verified, they utter the *same* words and the live sample is once again compared to the stored sample. More advanced systems are capable of sophisticated natural language processing and artificial intelligence sub-systems that would allow people to use *any* choice of words—in this case the words are less important than the characteristics of their voice. These systems are well suited to facilities and systems access control where the comparison is a 1:1 search—the system is comparing a known person against an existing template. It gets extremely complex when searching a sample against a database of potentially millions of voices.

The human voice has its disadvantages and problems develop when your voice changes between the enrollment and verification stages. Your voice varies substantially based on a variety of factors not the least of which is your state of health and situational deviations that alter your mood. If you get a cold the pitch and intonation change as well; if you are exited or upset the cadence or speed at which you normally speak will increase. And these changes can be enough to result in errors depending on the sensitivity of the system and spectrum of variation it can tolerate. There is always the potential that your voice could be recorded and played back and most systems cannot differentiate between a live voice and a recorded voice. The latter problem can be resolved using a second authentication method such as passwords or PINS that must be typed.

KEYSTROKE DYNAMICS

If quantum holograms sit at one extreme of the technology, then keystroke dynamics sit at the other, admittedly, less extreme. In a nutshell Keystroke Dynamics is a type of biometric measurement that verifies identity based on how they type at a keyboard. Originating with Morse code, the technology focuses on the characteristics of typing like pace, latency (time) between keystrokes, pressure, and the syncopated routine of a person's keystrokes against a stored profile. Even if an imposter acquired your password it would be much more difficult for them to precisely duplicate the sample pressure, and speed that you normally use. The only probable attack would come from covertly logging your keystrokes and playing back when they log in. It is certainly not of the better-known or trusted forms of authentication and would be better suited to lower-risk environments. It doesn't invalidate the concept—combining KD with a strong password policy would make it exponentially for casual attackers to gain access to a building or network.

FACIAL RECOGNITION

Prior to September 11th, 2oo1, few had ever heard of facial recognition but that has changed in the last four years as the controversial technology has begun to find itself in anywhere ranging from airports and malls to football stadiums…actually any closed circuit television system can be integrated with facial recognition systems. Facial recognition is not controversial because of *what* it is looking at; it is controversial because of *how* it is performed. While fingerprints and retina scans are intrinsically overt and visible, facial recognition is comparatively passive and covert—the public is rarely if ever informed that they are under surveillance nor are they able to provide any form of consent.

Quite understandable in an airport but it raises questions when this surveillance is used in the mall and other public venues. That was certainly the case when the Tampa Police Department installed facial recognition systems at Super Bowl XXXV—Tampa also installed 36 cameras in the Entertainment District of Tampa Bay and linked the cameras to a system called FaceIt with the premise that the system may be able to identify persons of interests. What they did not expect was the firestorm that ensued.

The human face is unbelievably complex and yet, somehow, humans have little difficulty in picking a person from a crowd while computers still struggle to achieve the same task; we have been doing this since the beginning of the human race while computers are just beginning to learn this complicated task. Faces may appear structurally the same but the peaks and depressions are unique to each individual and in the world of facial recognition those landmark features are referred to as nodal points that number approximately 80 and include the distance between your eyes, the width of your nose, the depth of the eye sockets, structure lines of the chin, and many others. When taken together, multiple points are converted into numerical codes and compared by software to determine if a person's facial "features" match a stored profile.

Capturing a human face is incredibly difficult. First of all, the system must be able to determine whether an object in a camera frame is a person's face in less than a second and decide whether to capture the image in multi-scale resolution that can switch resolutions once the image is captured. Following the capture step, the face in the image must be re-aligned before the other processing steps can be completed. If the person is turned away from the camera or perpendicular to the camera then the image cannot be processed—they must be looking indirectly at the camera at no more than a 30-40 degree angle. The alignment phase rotates the head and scales the image in terms of size and distance from the camera so that it will be comparable to the stored faces in the database. With a comparable image stored temporarily, the image is then converted into a unique face code using either Local Feature Analysis or other algorithms, which would then be compared against a database of stored codes.

The science of facial recognition is hardly perfect and is still susceptible to errors and deception—at this point an individual can obscure his or her appearance through glasses, wigs, and in more extreme circumstances, plastic surgery. What it is more disconcerting is the potential that facial recognition can threaten civil liberties—it does not take much of a leap to justify cameras at every intersection in the name of national security or public security or whatever mantra is currently in vogue. How long will it be before the police are equipped with mobile devices? How long before the picture on a driver's license is tied to facial recognition systems? There are still a lot of questions. There is no evidence, to the best of my knowledge that would suggest this ability has either been developed or that it would be deployed even if it was technically feasible. Then again, it took me less then ten minutes on the drive home from work to figure out how this system would function and there are people that do this work everyday—if I can figure it out then someone else already has.

IRIS AND RETINAL

Long thought the exclusive preserve of spy movies and intelligence agencies, retinal and Iris scanning have become the most accurate means of biometric authentication even though it is the most expensive (so far) and the least appealing to the public—most of us become rather skittish when anything is brought near the eyes and it remains one of the more invasive techniques. Retina scanning has been around for years and involves the analysis of the blood vessels at the back of the eye that is acquired using a low intensity light source through an optical coupler. Iris scanning is somewhat less intrusive but nevertheless entails scanning the rich colored ring of tissue that surrounds the pupil using a standard camera configuration that doesn't involve direct contact with the eye. I do not suspect either technique will find broad usage with the public due to the proximity to the eyes but you should know it does exist.

BIOMETRIC FUSION

It stands to reason that if one biometric measurement can provide so much potential, why not just use several of them to ensure a higher probability of identifying a person and increase the confidence level. Even if it is not common in the majority of installations, it is possible to find biometric fusion in the most secure of installations such as nuclear reactors, government agencies, and combat control centers on naval vessels. Biometric fusion centers on the capture and comparison of *multiple* biometric measurements like a fingerprint and retina scan—usually from two different body parts. The measurements are then scored and compared against an aggregate threshold—if the score surpasses the threshold then it is a positive match.

Overall, most of these technologies are not new but it is only recently that they have received more attention in the media; it remains to be seen when or if you will ever be subjected to fingerprinting but I'd venture it will happen within our lifetime. Banks and financial institutions will be the first to implement it followed by merchants. In the United States, as I mentioned, several supermarkets have already begun to market biometrics as a secure, convenient means of account management and it is only a matter of time before this proliferation broadens to other sectors. But before judging prematurely, this technology can be used for good as well as evil. I will get to some of the more common applications; fingerprint scanners can be used to make sure only the parents can take a child from a school or start a car. Fingerprints can be used to supplement cryptographic methods to secure medical records thus making the data available only to the holder of the fingerprint.

5.3 The Typical Biometric System

"They who would give up an essential liberty for temporary security, deserve neither liberty or security"

—Benjamin Franklin

Biometric systems can vary quite a bit in their methods and architecture as well as the underlying science but there are some strikingly similar features that can be generalized to provide a rudimentary model of the typical system. As you might expect there are many components and sub-systems that begin with the collection of a fingerprint or face print followed by an analysis and comparison to authenticate a person's identity.

DATA ACQUISITION

The first key sub-system involves the collection of signals from a person using some form of user interface and sensory equipment. Obviously the user interface must be very easy to use and not be unduly awkward that it would make it too difficult to contend with by the average individual. Every system requires signals or images captured from the person that are temporarily stored for further comparison. The data can reside on a single machine or be transmitted across a network depending on the size of the organization. A building access control system could run on a single machine whereas a network authentication system could require multiple locations and be distributed across countries and even continents. No matter how good the sensor equipment is, there is a good change that the incoming image might be distorted or "dirty"—perhaps a person accidentally smudged the sensor plate when providing a fingerprint. Whatever the cause, every image is sent to another component that refines the image and corrects source distortions before the comparison phase.

Once the signal has been acquired, cleansed, and stored, every biometric system requires a pattern matching sub-system that performs the feature extraction and pattern recognition of the stored template and the live image. There are all sorts of image extraction and pattern-matching algorithms though they tend to serve the same functions—extract common features from a finger or face and compare against a known image and determine if the two images (templates) are the same based on an acceptable threshold of variance. Even if the score was 80% (due to image processing anomalies) the system may score a positive match if the threshold is relatively flexible.

STANDARDS

In response to concerns that systems are easily connected and raising the prospect of a surveillance state, biometric vendors have been quick to argue that their systems cannot be connected due to a closed, proprietary architecture. But proprietary systems are a thing of the past and companies have recognized the value of conformity with international standards. It is conceivable that some systems are indeed closed in the sense that they use revolutionary algorithms or protocols but there is, nonetheless, an industry push towards uniform standards for the processing and exchange of biometric data. Specifically, from a development point of view, software engineering can be simplified and accelerated by a series of "interfaces" and libraries referred to as the BioAPI that can make the building of the applications as simple as using a library of pre-built functions. The point of these libraries is to eliminate the coding of lower level "primitive functions" so that developers (and businesses) can focus on developing the business logic. This allows organizations to compete more effectively by reducing costs and time to market.

Likewise, there is quite often the need to transfer biometric data between systems and this exchange can be facilitated by the Common Biometric Exchange File Format (CBEFF-NISTR-6529) that was developed by the Information Technology Laboratory of the National Institute of Standards & Technology. The aim of the work was to create a "technology-neutral" or technology-blind format that could be used with any biometric system. In so doing the standard does not show bias towards any type of technology or biometric measurement, it simply provides a vehicle for the sharing of information. It would still be up to vendors, organizations, and governments, to ensure they are using the same algorithms.

Though this represents a very simplistic view of the typical biometric system, it does suffice to provide a better understanding of *how* complicated these systems are and the complexity of the science involved. This is hardly at par with programming an Excel spreadsheet. But it does go to the degree in which the field has advanced and will continue to advance in the next 10 years. The use of standards allows vendors to develop an ever-increasing number of biometric applications and it is an important passage in the history of biometric sciences.

5.4 Applications of Biometrics

"The body says what words cannot"

—Martha Graham (1893–1991)

A majority of us that have worked in the corporate sector are already accustomed to security in the form of building swipe cards. But ever so slowly these cards are gradually being replaced or supplemented by fingerprint or palm scanners. From prisons and military installations to banks and supermarkets, the measure of the human body is going to either replace physical tokens or support them as a second line of defense. From a simple single-system facility control systems to highly complex transnational computing environments, the need to have a high degree of confidence in the identity of employees and customers is changing the face of industrial security and society.

To see how biometrics fits into the security paradigm, you must first become acquainted with the three types of security. The first, referred to, as Something You Know (SYK) is something you are already familiar with—passwords and PIN numbers. The premise of SYK is that only an authenticated user will know a secret word and access is granted on the basis of this knowledge. Of course, password theft is exceptionally easy and anyone in possession of a password automatically assumes the identity of the target. If I capture a password and login with your credentials and send mail through your account then there is nothing to say it wasn't you that sent that e-mail. The second type of security is known as Something You Have (SYH) and is best characterized by your building swipe cards and bank cards. Here too it is assumed that the person in possession of the card is the right person—if an imposter acquires the card then they will accrue the same privileges as the card owner. Before getting to the last type, an improvement in security would be to combine SYK and SYH which are referred to as multi-factor authentication—it is much harder to impersonate a target if the imposter must acquire their card *and* break their password. The last type of authentication, and the focus of this chapter, is known as biometrics or Something You Are (SYA). Intelligence agencies and movies notwithstanding, these are the strongest types of authentication and almost impossible to impersonate.

Using biometrics in buildings or to secure networks solves a lot of the authentication problems with SYK and SYH security, especially in ultra-secure installations like nuclear power plants and research centers. There is nothing to carry or nothing to remember, it is just a matter of placing a finger on a small pad and the door either opens or it doesn't—that is a rather rudimentary description but essentially that is what biometrics does. The following describes some of the better-known uses of biometrics—it is not complete but it does provide a quick and dirty overview of how it is used.

COMBATTING FRAUD IN SOCIAL SERVICES

Social services are another area where biometrics is being deployed to combat fraud. Welfare and benefits fraud have reached epidemic proportions and cost billions in losses as well as depriving those who are legitimately in need of assistance. These losses are making it much easier to justify biometric authentication and some states including Connecticut have legislated that welfare recipients must be imaged and it is a fair assumption that other states will follow in their footsteps.

There are several common fraud patterns but the most prevalent scenario involves a person either applying for benefits under more than one name or collecting benefits from more than one jurisdiction—it is equally possible that the person could be claiming under multiple names in multiple places. Using advanced authentication methods, be it fingerprinting or whatever the case may be, has the potential to prevent the crime *before* it occurs. Traditionally officials must conduct name searches that are based on the assumption that a person's documentation is valid—if the birth certificate is false or the person has assumed an identity then a name search becomes futile and creates an environment ripe for fraud. Biometric authentication would virtually eliminate these kinds of frauds and may well serve as a deterrent.

This kind of "eligibility protection scheme" is not without its opponents and the arguments are valid. In addition to the violation of privacy, there is the concern that fingerprints would be shared with law enforcement which should *never* be permitted without a court order or if a fraud has been committed. To wantonly and indiscriminately share fingerprints with law enforcement infers that all recipients are *guilty on the basis of poverty*. It will be a lot easier to support these programs if there was no sharing.

PRISONS AND CORRECTIONS

It would be hard to find fault with the deployment of biometrics in prisons; unlike the civilian population that is entitled to guaranteed freedoms under the constitution, inmates of a correctional facility have a diminished expectation of privacy within the confines of a penal institution and, for what it is worth, fingerprinting likely pales in comparison to the other indignities that inmates must ensure. But what benefit does biometric surveillance offer to prisons when the population is already confined as it is?

In any type correctional facility of any type or size, the most important objective is the controlled movement and accountability of both inmates and staff at any given time or location. The challenge of movement is to control the whereabouts of all staff to specifically prevent the potential for hostage incidents and violent acts. Likewise, the challenge of accountability is to positively determine the identity of each person. Biometrics can effectively provide solutions to both challenges and it has proven very successful thus far.

To ensure accountability some facilities use biometrics to positively identify an inmate when they arrive at a facility and when they leave. If a person is granted parole, it is critical that authorities can be assured, unequivocally, that it is the same person. In Pennsylvania, inmates are required to submit Iris scans when they are admitted and processed; when they are scheduled for release, they are scanned again to make sure another inmate hasn't taken their place. The same technological solutions can be extended to the movement of people within the facility; fingerprints or a palm print are required to move from one area of a building to another and protect everyone in the institution. This type of monitoring protects everyone—including prisoners from each other; no one from the general population would be able to get near other prisoners in protective custody. It might even play a pivotal role in hostage situations—a facial recognition system from NEXUS which already exists that will assist officers in identifying which inmates were responsible for a riot or hostage taking.

Ultimately the application of biometrics in corrections is an appropriate application of technology to solve a social problem and I personally have no misgivings with respect to this particular use of biometrics; that live scan fingerprints can mitigate or eliminate the risk of physical harm to inmates and staff is sufficiently justified in my opinion and ensures the security of a community.

LAW ENFORCEMENT & NATIONAL SECURITY

Regardless of the statutory and constitutional constraints on the use of biometrics, it is still one of the most pervasive applications in this field and it is deserving of attention. The applications of biometrics in both law enforcement and national security are very closely related and based on the obligation of government to protect the population from a variety of threats. Insofar as law enforcement is concerned, the possibility of identifying suspects in real time, and more importantly, in the field must be very alluring to police forces. There is considerable disparity between police forces in different regions of the United States; while some precincts are being equipped with the latest in Blackberry PDA's, other precincts are still running on antiquated PC's so it will be some time before this becomes a regular tool for police.

The national security position is much different and biometrics can offer both defensive and offensive capabilities that were probably the silent thoughts of an active imagination. It is difficult to know, with any degree of certainty, what capabilities the intelligence community really has, besides a little bit of playful speculation. Defensively, the national security objective is to identify threats before they are able to enter the country and cause harm; the airports may be much more secure but there are still land routes and marine points of entry that would be much easier to penetrate. Hypothetically, authorities might be able to identify a print already in their systems and red flag suspects before they ever cross into U.S. airspace. That assumes, of course, that U.S. intelligence has positively identified the target and have acquired their prints but it is still feasible and quite a few countries, including Canada, are moving towards a biometric passport so it is only a matter of time before all countries adopt similar standards. As a general rule, the United States government is permitted under the Constitution to do whatever is necessary to protect its borders and the Border Exception Rule of the Fourth Amendment allow for all sorts of invasive surveillance at American points of entry. If U.S. Customs can search through a person's luggage without cause, you can be quite sure that the use of facial recognition and fingerprinting falls under the same right of government to protect its borders.

Offensively there is research being conducted at the Defense Advanced Research Project Agency (DARPA) that would allow the intelligence community to identify a person from a distance based not only their face but also how they walk and move (Gait Analysis). Assuming they can determine that an unknown person was a terrorist, it is technically within reason that the threat can be neutralized at a distance. The current system, at last reading, is capable of identifying a person at 500 feet under varying weather conditions and it would be a fair prediction that the range and accuracy will improve in due course with the presumed objective of recognizing a person from aerial or satellite based platforms.

The danger of any kind of law enforcement surveillance is the potential that such acts will constitute a violation of search and seizure and due process; it also threatens to turn upside down the accepted principle of innocent until proven guilty. This has become a timeless debate that I will discuss in more detail in Chapter Six but it bears mentioning that democracy is grounded on freedom and to deprive Americans of their freedom is not the so-called American way no matter the reasoning. The national security angle raises just as many questions and the key concern is that terrorism will be used as an excuse for intelligence agencies to turn their muscle on the very people they are supposed to protect.

BIOMETRICS IN THE WORK PLACE

It is much easier to justify the use of biometrics in an airport where there is a diminished expectation of privacy but using biometrics in the workplace is a different matter and should trigger an uneasy feeling given the potential for widespread abuses; that doesn't mean that every organization would use it to the detriment of their staff but the fact remains, and court decisions have proven, that there is a history of abuses. Not every application of biometrics in the workplace is abusive and management has the right to monitor the arrival and departure of their employees and time attendance systems are already being used to this end. It is also reasonable for organizations to track staff as they enter and leave secured, sensitive areas of the company such as laboratories or vaults or computer facilities and I have no qualms with the right of employers to secure their facilities and monitor the performance of their employees within reasonable limits.

The fear, and it is a reasonable concern, is that so-called time tracking will mutate to track all of the movements of a person within a facility—at some point someone will get the "bright idea" that the system can track *every* movement—to the bathroom, down the hall, five minutes at a colleagues office, ten minutes at the cafeteria etc. Quite frankly the objective of these strategies—to improve productivity is flawed and predicated on questionable management practices–draconian measures reduce productivity and significantly reduce staff retention.

I am not opposed to biometrics in the workplace with the proviso that it be applied sparingly and judiciously—employees must be educated and the reasons for surveillance must be clearly and unambiguously articulated. Another problem with biometrics is the legal abyss that could follow—just because the Employee Handbook says so does not make it so, nor does it make it legal and every management team would be advised, in the strongest possible terms, to seek counsel before implementing biometrics in the workplace or face the possibility of criminal charges, human rights violations, and civil suits.

There is also indication that biometrics is making headway into the mainstream with fingerprint scanners now available in some shopping venues; rather than using a credit card, customers use their fingerprint to perform a validation and debit their account for whatever they purchased. If there was ever a case where convenience had run roughshod over privacy, this is it. It is one thing to demand an Iris scan to provide security on an Aircraft Carrier and another to demand a fingerprint to buy a coke.

5.5 Boundaries of Biometrics

"In Cyberspace, the First Amendment is a local ordinance"

—John Perry Barlow

If biometrics present as much of a threat to liberty as it has been suggested then why haven't more people heard about it? Perhaps you have, but there is the question of how it was presented; if the only exposure to biometrics is through television and an obsession with new toys then there is little likelihood that the privacy questions will get any airtime. Based on my own observations there is a profound public apathy towards privacy, particularly when there is no tangible immediate threat.

It would be a grievous mistake to think that public attitudes will remain constant over time; to the contrary, government and private sector organizations ought to exercise caution lest the winds of fate change against them; that very same sense of public indifference could very quickly change to anger if the public comes to grips with the erosion of privacy. I cannot predict what catalytic factor will swing the pendulum of public sentiment—it could be a single event or a series of events over time. Maybe it is because the public does not know what questions to ask or that the media have failed to truly engage in what we once remembered as investigative journalism.

The most important question to be asked is whether a security technology is truly effective and whether the civil costs justify the intrusion on upon the privacy of the population. What is the public willing to sacrifice in exchange for a meaningful improvement in Security? What level of surveillance would be acceptable to any given community if it meant a reasonable reduction in the level of crime? Would you allow a camera in your home in exchange for no crime? Or would you be willing to take a minimal risk in exchange for your privacy? That is really what you are being asked.

I am sure it will come as no surprise that there has historically been a rather militant reaction against biometrics from the privacy profession while the general public has been relatively unaware of its existence barring the odd news. That is changing as more government agencies and a small number of private companies begin to deploy biometrics for building security and other applications. The objections tend to focus on the legal issues and weaknesses of the technology while little is ever mentioned of its potential benefits.

Admittedly, the broader proliferation of biometrics could, potentially, result in drastic changes to our culture that would be impossible to predict today. There is also the very reasonable fear that this could reinforce the growing sense that we are approaching the dawn of a true surveillance state. The fact is the original Brandeis doctrine on the Right to Privacy is just as true today as it was over 100 years ago—even with changes in technology, people haven't changed that much.

At some philosophical level, this particular technology runs the risk of robbing a society's right to self-determination and even the freedom of movement—we cannot make decisions effectively or move freely if there is always someone wanting to know precisely where we are at every waking moment of the day—it is akin to the micro-manager where you work (every company has at least one); the more they hang over a person's shoulder, the less effective that person becomes.

Thankfully, there are both a significant body of constitutional jurisprudence and myriad regulatory regimes coupled with the threat of civil litigation that serve to control or mitigate the aggressive proliferation of biometrics. Society is governed by laws, which define our behavior and what fits within the realm of acceptable behavior not only by people but by organizations as well. And it is that threat of penalty that will help to curve the appetite to control others—it won't prevent it but it will keep a lid on it. A technocrat will always find a way to justify the use of technology, more often than not without an adequate peer review or environmental assessment. Progress for the sake of progress is not progress.

And yet it is equally hoped that you do not blindly condemn biometrics. There are many legitimate for deploying it to truly improve security and safety in some environments. And there is reason to believe that biometrics could even serve to enhance privacy rather than erode it. New forms of documentation like passports can be engineered to provide additional layers of security that would make a passport entirely unforgettable. Biometrics could also be used, with the right safeguards, to allow parents to track their children's whereabouts using satellites—the biometrics would simply be used to authenticate the parents. To understand an application requires that we understand how it is going to be used and what safeguards have been designed into it to protect the population against abuses. Who has access to the data, how will be it stored and shared with other parties? These are the kinds of questions that have to be asked long before the first line of code is ever contemplated.

And then there are those applications that were discussed earlier which are much more difficult to oppose—I still cannot find any meaningful objections against using biometric authentication in nuclear power plants, defense facilities, and prisons—places, in other words, that demand a higher level of security and therefore a diminished expectation of privacy. Others like applications for the workplace require a much more rigid review and regulatory approach to ensure they are limited if not banned entirely.

As it is often the case, technology may indeed provide improvements to our standard of living, but it is the laws, policies, and human oversight that are needed to ensure that there are adequate checks and balances and that the pendulum doesn't swing too far either way.

Biometrics is clearly one of the most serious surveillance technologies that could change the face of civilization and curtail, if not permanently damage, privacy and fundamental civil rights unless steps are not taken to control the proliferation of applications in both government and the private sector. Self-regulation has historically proven ineffective and analogous to the wolves guarding the hen house—voluntary compliance alone will not work. A more robust approach would entail a multi-layered legal enforcement regime as follows:

- *The Constitution must be amended to account for changes in technology or worded in a manner that would give full constitutional protection to privacy regardless of the technology. This is the baseline above which other more specific controls can be layered. Furthermore, it should be applicable as much to the private sector as it is to government abuses.*

- *An appropriate and comprehensive federal and state statutory framework that would further limit the use of biometrics.*

- *A voluntary industry code of conduct that exceeds the standards rather than just meeting them. It is time the industry show it truly appreciates the privacy implications of their inventions.*

- *The right of states to augment federal law by enacting an even stricter regulatory regime that would be in line with the community standards of the state.*

- *The last resort is civil litigation under which violations of privacy may become the subject of civil lawsuits and which have already occurred in the courts.*

It is unfortunate that organizations and governments do not recognize that every search of the human body, regardless of the rationale, constitutes a fundamental invasion of the human body—and that those accountable might not be so vociferous in their support if they were the targets of such indiscretions. But until that time the best recourse is a powerful legal framework that will provide a safety net until we come around to realizing the value of privacy. Biometrics, if it was harnessed and properly controlled, has the potential to deliver many benefits including improvements in security and it might even be turned around, as I've pointed out earlier, to actually protect privacy. Until then, time will tell whether I am right or wrong.

6

The Surveillance State

6.1 The American Surveillance State

"A satellite has no conscience."

—Edward R. Murrow (1908–1965)

While there is little doubt, particularly amongst civil libertarians, that the United States is dangerously approaching the abyss of a surveillance state, and there is ample evidence contained in this book and many others to support that supposition, it begs the question whether this state of affairs is just an aberration in our social development or whether it is the inevitable consequence of technology and progress. Technology has advanced to the extent that surveillance that was impossible twenty years ago is now considered very common. There are cameras that increasingly show up at intersections and in the malls, automated cameras take pictures of our license plates and send equally automated bills for minor traffic infractions or using toll roads. Technology is changing how we live and those watching how we live.

And yet there is so much more the public does not understand about the culture they live in, nor than they would the technologies that watch us. If you can legally buy spy equipment at a store, can you imagine what equipment is used in the course of watching everyone else? If you can spend $200 hundred dollars on a web-cam to watch your living room, can you just imagine what $200 billion can get your government? A bigger camera! A much, much bigger camera. The 21st century, though still in its infancy, is already witness to a dizzying array of new surveillance technologies including massive databases, satellites, cellular tracking, RFID chipping and many more. Indeed many technologies that were never meant for surveillance have been effectively co-opted for the express purpose of watching us.

The fact remains that privacy invasive technologies that can reach into the very depths of your existence coupled with the government's insatiable desire for more surveillance, especially since September 11th, are changing the very fabric of American Culture, the constitution, and the civil liberties that are supposed to define one of the most democratic nations in the world. . Somewhere, someone has the potential to read your mail, see the library books you checked out, the videos you've rented, where you went, what you bought, and who you called.

A surveillance state, though differing from a classical police state, focuses on the identification, classification, and recording of information about its citizens through increasingly technological means. These collection efforts are often coupled with efforts to confuse the public through misinformation that ultimately shapes public opinion and attitudes towards government. It further seeks to shield the less than palatable elements of the state through restrictive secrecy guidelines that prevent the disclosure of material facts.

Lest anyone rush to judgment, I must make it abundantly clear that I do not fall into the camp of conspiracy theorists and give little credence to their delusions—that does not mean strange things don't happen but I try instead to focus on the science and evidence. My duty is to present to you a snapshot of the systems that currently exist so that you can come to your *own* conclusions. That was my aspiration when I first considered writing the book and it remains a central theme—give you the information you need to effect choice. In this case, there is little expectation that these systems and their sponsors will simply disappear or that abolition of the systems can be affected by one author. I am not even opposed to an effective domestic counter intelligence infrastructure if only the agencies responsible for said duties would conduct themselves intelligently. That means focusing on those that present a real threat.

There is still sufficient factual evidence, often from the government itself, that there are some really dark and ominous signs that our civil liberties are constantly under assault. The *fact* remains that there *are* indeed systems like Carnivore and Echelon, there *are* laws that that allow law enforcement to arrest or search you without probable cause, and there are exemptions to the constitution that give the government extremely broad police powers. They are not the things of a distant future, nor the ranting of a conspiratorial mind—they are printed, though often obscured, in our laws.

My interest for bringing the subject to light bears relevance to the book's focus and to ensure that the collective national defense infrastructure is directed outwards towards enemies not inwards towards those it was supposed to protect. Contrary to many in the privacy field, I am an unrepentant proponent of national security and think we have done a great disservice to the armed services, particularly in Canada where the Liberal government has all but decimated the Canadian Armed Forces. Despite that support I am not blind to documented abuses and I am certainly not the only one to have grave reservations about the direction and focus of our intelligence services—they ought to focus on those they truly see as national security threats rather than developing systems that assume everyone is a threat—and that would, on occasion, appear to be the direction of the systems I will be discussing.

The current wars in both Afghanistan and Iraq coupled with 9/11 and ongoing terrorist threats have also given rise to new laws like the U.S. Patriot Act, amongst others, that are set to undo what the constitution enshrines and provide the government and law enforcement with unfathomable powers including the ability to see what books you have read and what movies you've seen—and I would sincerely doubt if that is entirely restricted to the United States, so it would be best not to be too smug. U.S. allies have similar acts (though often not quite as extreme) that permit similar powers.

Despite this, and there are hundreds of thousands of pages on the subject, I have found in my discussions with others not only a lack of understanding but, worse still, a seemingly lack of concern or indifference. Perhaps, we have been conditioned to believe in the mantra that we have nothing to hide, therefore we have nothing to fear—surely the surveillance state will focus on those deserving of its attentions and not Mr. and Mrs. Smith—unfortunately it is not so simple.

Authors often risk another backlash from government and the press when exposing national security and privacy issues. Such exposes tend to cause a rather "vocal" reaction from the government and their proxies, most notably the insinuation that to oppose national security measures is naïve at best and unpatriotic at worst. They are accused of being uninformed, or unpatriotic, or delusional with thoughts of conspiracy. It is not unpatriotic to question and challenge and criticize the organizations and systems that were developed with the sole purpose of watching the public with the knowledge that these systems are often in violation of the constitution. As for being uninformed, thank those that have sought to obfuscate the truth—then again they ought to know by now that eventually the truth finds the light of day.

The fact is there are indeed massive national and state surveillance systems that were designed and implemented for the purpose of watching Americans (or Canadians, or Britons…or…). These systems are integrated with ultra-large scale databases that are capable of storing everything from where you went to school (and your grades, sorry about the "F" you got in Gym Class), the medications you take, the books you read, the movies you watch, what tickets you have gotten for speeding, all of your ATM transactions, and so on. Remember what I said about data mining—the goal is to create massive profiles, and that applies as much to government as it does to the private sector.

But how effective are these systems at catching the people they were originally designed to capture in the first place? If a CCTV camera catches me speeding or jaywalking, or a neural network determines I deducted too many expenses on my tax return, I have no doubt they will catch up with me; but can the same be said for the terrorist that slips through the facial recognition system at the airport. So these systems, barring evidence to the contrary, appear to do little to improve public safety and security but cost us dearly in the freedoms that are diminished.

One of the principle concerns of the ACLU has always been the disconnect between what the systems are supposed to achieve versus what we are losing in exchange—if these systems offer a sense of false security than we are no better off. Furthermore, violations of the Fourth Amendment on search and seizure are compounded by violations of the Fifth Amendment with respect to due process since these actions are often take without our consent and there is no legal recourse to correct mistakes.

How does any society find a sense of balance between the need for a given state of security and surveillance against civil liberties? Weaken those civil liberties too much and western democracies become no better than the dictatorships and regimes that we've gone to war with under the guise of saving the oppressed. You cannot tell a neighbor to clean their backyard if yours is no better. I am by no means opposed to the use of surveillance systems and would go so far as to say the government also has rights enshrined in the constitution that many are unaware of. What concerns me is that elements within government would abuse those provisions to further their own political aims or pursue a police state under the umbrella of anti-terrorism. Just as there are those that want to destroy our world because of their religious beliefs, there are also many law enforcement and intelligence officials that would seek to undo protections in the name of security—permanently.

For that reason there must be a means of returning to a sense of social balance where law enforcement can perform their duties in our interest while safeguarding our civil liberties as much as possible. That is why we require oversight and public scrutiny in all such matters—the light of public scrutiny does much to prevent the dark and sinister temptations of the few. It must also be determined, in an objective unemotional forum, whether new surveillance and restrictions are actually beneficial, effective, and somehow able to provide real improvements in security. Throwing grandma to the ground because she forgot to put her nail file through the X-Ray machine is no way to inspire public confidence.

Although this chapter focuses on the technologies that are used in the pursuit of state surveillance, technologies are just weapons in the hands of those that do not see the Constitution through the same optics as the rest of the population. And rather than openly debate the security problems facing Americans or Canadians, governments have opted to enact draconian legislation that will do little to stem the tide of terrorism or any other crime. For that matter every major department in the government of the United States has put forth their own brand on security that is having a chilling effect on freedoms and privacy.

Attention may be focused on the FBI and the Department of Justice after the Patriot Act became the lighting rod for those opposed to surveillance, but lets not lose sight of the other changes that have slipped under the radar. The Department of Transportation has its own "no-fly" lists of those either suspected of terrorist or criminal activity or that has openly criticized the administration. The Department of Defense has been developing a large number of systems that are partially discussed in this chapter such as Total Information Awareness. And these are just the ones we're aware of. How systems exist that have yet to see the light of day.

All of this would be fine if these initiatives were really directed at terrorists but history is replete with examples of intelligence services being turned on their own populations either due to national paranoia or the wishes of a cynical administration. If there is any doubt, it would be wise to recall how both Nixon and Johnson used the FBI to conduct illegal surveillance on political opponents and it raises the issue of what constitutes a threat to the government. Does an overt act of terrorism constitute the threshold, or is writing an unfavorable book critical of the government enough to warrant that dubious distinction? I don't have the answer to that but I would be amused to see my own files.

I wouldn't want to give you the impression this is a complete or exhaustive treatment on the topic of surveillance states or the technology that is being used;

it is, like much of the book, a humble snapshot of the major systems that you may have heard of with a little more detail to make the point they exist and that your privacy is being eroded by their very existence. I am nonetheless not as worried about the technological advances, for they will come and go, but rather the dent it puts on freedom of expression and movement. Next time a police cruiser is driving behind you, ask yourself whether you just dropped the gas a little. Now count the number of times you check your rearview mirrors—more than you normally do.

6.2 The FBI Carnivore System

> *"And so, my fellow Americans: ask not what your country can do for you—ask what you can do for your country. My fellow citizens of the world: ask not what America will do for you, but what together we can do for the freedom of man."*

> —John F. Kennedy (1917–1963),
> Inaugural address, January 20, 1961

While my remarks prior and following may give the impression that I am against any type of surveillance, this is not the case. And it is imperative that readers always keep in mind that, unfortunately, serious crimes are committed every day and many of those crimes rely upon the national telecommunications or internetworking infrastructure to conduct planning, collaborate, and execute crimes. As well the more sophisticated of criminal activities include but are not limited to organized crime, white collar crime, narcotics, terrorism, smuggling, espionage, and so on.

That does not diminish the impact of robbery or assaults, it is just that these crimes are often larger in scope and require a different kind of surveillance. So in fairness to law enforcement and the government agencies that are mandated to prevent these crimes, they must be able to evolve with the times and leverage technology to keep up with organized crime. It is counterproductive to put *unnecessary* constraints on police forces to satisfy the paranoid delusions of the few that never seem to find anything good about law enforcement. Those that execute these crimes seek to deprive people of their rightful property, financial security, let alone their sense of trust, safety, dignity, and in many cases the lives that are either directly or indirectly lost. This isn't a carte blanche endorsement for the police to wantonly abuse the constitution or civil liberties—it is simply the grim reality that the police must be able to use whatever is available to prosecute these cases.

That said, the first major system of this chapter, the only one that has garnered significant media attention, is the FBI Carnivore System that was developed to capture e-mail and which is targeted at specific individuals and organizations. Unlike some of the other initiatives that are to be discussed later in the chapter, the Carnivore system is not classified as intensely as some of the others like TIA and Capps due to sufficient evidence to indicate that the FBI has taken reasonable precautions against systemic abuses and the extent of oversight is higher with this program. Any surveillance system of this magnitude is capable of wide spread systemic abuse and it is absolutely essential that it be properly managed, controlled and supervised by the courts; nothing short off intensive operational scrutiny and transparent judicial mechanisms are necessary to ensure that abuses are the exception instead of the norm.

So what exactly is Carnivore? Put in the simplest terms, Carnivore is a special purpose computer that serves as a "packet sniffer" that can surgically intercept specific types of Internet traffic coming from or going to a specific person or group that has been identified in a *court order*. Once *authority* is granted by the courts and a senior official with the Department of Justice, the FBI technicians and agents work with the technical teams of an Internet Service Provider to install the system on the ISP's network, typically off their router(s). The team then configures a set of "filters" that define what type of traffic and who they want to trace, which could be anything from e-mail to web addresses and chat conversations. There are supplementary components to Carnivore like the DragonWare Suite that allow the FBI to "reconstruct" the web sites a suspect has visited. The importance of these filters lies in their ability to limit what the FBI can and cannot capture. In fact, according to the FBI, the filters are only supposed to capture the data and addressing information but not the message itself, though I suspect this happens more than they would be willing to acknowledge. First of all, there is a strong possibility that they already have access to the content of the messages. Secondly, data and addressing information, in and of itself, does not definitively define a criminal act. Thirdly, if wiretaps can capture the content of a phone conversation then an e-mail tap can, if permitted, capture the body of e-mail. Once the system is configured, Carnivore begins to capture all of the specified traffic when the users log in to their ISP.

Every system is going to have its share of technical and operational challenges, and Carnivore is not immune to problems and based on available data, accidentally capturing unintended information from other people or capturing traffic not authorized by the court is still an issue. With older types of intercepts, such as the traditional wiretap on a phone, the subject is relatively segregated and the probability of capturing traffic from someone else was negligible—this is not the case with Internet traffic where all traffic comes through the same "choke points". Furthermore, unique identifiers like the ESN's used in cell phones make it easier to restrict the intercept so that it is only capturing conversations from the intended target.

It is possible that if Carnivore was incorrectly configured it could theoretically capture all of the Internet traffic of an ISP—a proverbial fishing expedition, and these accusations have been levied against the FBI. The FBI says this is not possible but I've configured enough packet filters to know it wouldn't take much to capture everything. Of course capturing all of that traffic would be overwhelming, something I and many engineers have learned the hard way. To be somewhat more reasonable, it is expected that mistakes are going happen and an agent might accidentally capture the wrong messages—there is going to always be an element of human error and stricter standards must be in place to make sure unintended traffic is either blocked or destroyed if it is intercepted.

Because of these concerns and a lot of pressure from privacy advocates, the Attorney General directed the Justice Management Division to conduct an Independent Technical Review to determine if Carnivore presented undue risks to citizens or ISP's and whether the system could be properly configured to prevent accidental mishaps. Then again your idea of "independent and objective" may differ from theirs and you would have expected the FBI to be more sensitive to perceptions of impropriety. So they naturally contracted the Illinois Institute of Technology Research under the Illinois Institute of Technology, the members of which also happened to have ties to the National Security Agency and several defense contracts not to mention a former advisor to President Clinton and the Science Applications International Corporation (SAIC), which in turn is the ISP for the FBI. Seems rather cozy does it not?

Now that I've covered some of the fundamental technical issues, it is time to examine the legal and constitutional questions. On the bright side, comparatively speaking, the onus of government to prove its case sits far higher than it does with systems that fall under the umbrella of national security and do not demand the same requirements of probable cause. In the case of Carnivore, the use of the system is covered by the Constitution and both civil and criminal penalties for its misuse do exist. Not only must the FBI obtain a court order in which probable cause must be proven, it must also seek approval from the Department of Justice.

It is in my opinion, based on extensive research of publicly available documentation, cases, and testimony that the FBI and its legal counsel has been troubled by aspects of potential abuse and have petitioned guidance in its use. It also appears that, in conjunction with this guidance, they have implemented very stringent rules and penalties that govern the use of the system by the technicians and agents. That doesn't mean that mistakes are not made or that some officers might overstep the boundaries of the law, but all outward appearances suggest that the FBI has tried to keep it as clean as possible.

Considering the some of the more draconian laws in existence, Carnivore is still one of the better-managed systems. Still, Internet service providers and telecommunications carriers are bound by the rules enacted by government and that includes the demand that all carriers cooperate fully in the execution of surveillance taps. As a prime example, the Communications Assistance For Law Enforcement Act of 1994 (CALEA) requires that assistance be given to law enforcement agencies in the execution of wiretaps and traces.

The "scary" law falls under Title III of the Foreign Intelligence Surveillance Act of 1978 (50 U.S.C. 1801) that makes CALEA look like a walk in the park. It suffices that anyone that is the subject of a FISA Warrant would do well to crawl under a rock. A warrant issued under FISA allows the police or government agencies to search, wiretap, and photograph a suspect without a public warrant or probable cause. In 1995, President Clinton *extended* FISA under Executive Order 12979 to include physical searches of a suspect that has been deemed a national security threat by the president or designated authorities such as the Secretary of Defense or the Attorney General. That hundreds of FISA warrants are issued every year infers that there is justifiable concern that there are a lot of people under surveillance without probable cause.

Despite the dark and ominous tone, the Carnivore System, at least when compared to other government surveillance initiatives, is much more open and transparent and the level of judicial oversight offers a sense the system is being observed unlike many surveillance efforts. And the FBI, to its credit, has been much more cooperative and transparent than many of its other agencies you will see shortly. Unlike national security systems, the benefit of Carnivore is that the evidence must eventually be presented to the courts. If abuses occur then it is well within reason to expect that the defense attorneys are going to find these procedural and operational problems in the course of their discovery phase of a trial. At the very least, the use of Carnivore, in a majority of cases, must conform to the law if the evidence is to be accepted the courts.

6.3 CAPPS: Computer Assisted Passenger Pre-Screening System

"Liberty is not a means to a higher political end. It is itself the highest political end."

—Lord Emerich Edward Dalberg Acton (1834–1902)

When it comes to privacy, the airport is one of the few exceptions to the rules and every passenger has come to terms with the diminished expectation of privacy that is to be expected when flying. Beyond the usual annoyances that are expected when traveling anywhere by aircraft, a majority of travelers give scant attention to the amount of surveillance and scrutiny as they pass through an airport; and considering the number of hijackings that have occurred since 1960, anyone would be hard pressed to find fault with this increased vigilance—indeed we *demand* safe air travel and most are willing to accept that increased safety requires increased security. It isn't just hijackings, there are also narcotics smuggling and fugitives crossing international boundaries that keep security officials awake at night.

So ingrained is the protection of America's borders, the Fourth Amendment on search and seizure has an exemption known as the Border Exception Rule that effectively means an airport is a security free-for-all when it comes to surveillance. That in and of itself isn't that surprising. What is interesting is the difference in strategies to security and the underlying attitudes towards travelers entering the country that one finds in traveling to other countries. While many countries have adapted universal screening and the use of human assets, the United States tends towards technological solutions.

In 1996 Vice President Al Gore was chairing the Presidential Commission on Aviation Safety and Security that was tasked with studying different approaches to improve overall aviation security and had focused on universal screening techniques that are used throughout the world. But the aviation industry lobbied aggressively and effectively against universal screening because that type of screening would slow check-in times, require people to arrive much earlier, and increase the time required to push aircraft through key hubs—in short it would cost a lot of money and also cost the industry in reduced revenues as people sought alternate means of transportation. This in the mind of the airline industry wasn't an option.

To delay or mitigate the possibility of universal screening, the aviation industry introduced a system called the Computer Assisted Passenger Pre-Screening System (CAPPS) that would collect passenger information in what is referred to as a Passenger Name Record (PNR) that contains over 39 fields of personal information. The commission elected to adopt the system as an *interim* measure though it is difficult to determine the amount of information this system really collects and how effective it is. This was supposed to be an temporary move until universal screening could be implemented but the recommendations were never put in place and CAPPS, rather than being mothballed, has evolved into CAPPS II.

Perhaps North Americans have been unduly spoiled by the lax security compliments of the airline industry; the fact remains that universal screening is the standard practiced around the world in which all passengers are thoroughly searched and questioned according to an established doctrine that is applied equally to all. The point is no one goes through the checkpoint without his or her bags checked and a determination made that the individual does not present a security risk. Universal screening eliminates charges of discriminatory practices since there are no exceptions and everyone is treated the same—it would not matter what country, religion, or race, everyone goes through the same routine. This more of an ideal than a reality, and it remains that classifying passengers based on their ethnicity is part and parcel of security screening whether anyone likes it or not. It may clash with contemporary sensibilities of equality but ethnic profiling is much more common than the government or airlines will ever admit to. But if stringent security measures and universal screening is the accepted practice in other countries then why has the United States been too reluctant to adopt the same practices? There are a variety of reasons not the least, of which are complacency, arrogance, and even greed. There is also a tendency to rely on technology instead of people that is attributed to the forces of greed. North American airlines have yet to accept that a well-trained and experienced security staff will outmatch any computer.

Have you ever seen an over-priced, overly complex system that government didn't like? In the case of CAPPS (and CAPPS II), the government became a quick convert and sought even more opportunities to mine even more information about passengers. By 1998 the U.S. government provided a profiling algorithm that would classify every passenger to determine whether they presented a threat (based on a set of criteria not available to the public) and then "grade" that passenger based on a scheme along the lines of green for trusted passengers, yellow for further screening and red which could ban you from any flight and prohibit you from traveling. There is also evidence to suspect that the United States have attempted, with varying degrees of success, to coerce foreign airlines and countries, into providing detailed information about passengers that bears no relevance to their travels—the implied threat has been that either they provide what is demanded or be denied entry into U.S. airspace.

As a general rule I am not opposed CAPPS or similar forms of aviation surveillance and the diminished expectation of privacy is a small price to pay to travel. But, so far, there is no published evidence to suggest that these systems provide measurable improvements in security and there is the risk that these systems may arbitrarily flag a person as a flight risk without the benefit of legal recourse. Furthermore any kind of recourse would be cold comfort when your vacation has been destroyed.

The TSA has acknowledged that 4% of the flying public may be accidentally flagged as a risk, which means there is a 4% probability that your flight could be disrupted, delayed, or cancelled completely. Chances are, those that are sophisticated enough to have alternative passports will pass through the gates unmolested while the grandmother from Florida gets hauled into a small room with a dim light.

Whether the billions that are spent on biometrics or next generation passports or information systems like CAPPS will actually prevent or reduce the security risks of flying remains an open question. In a telling COMPUTERWORLD interview with Isaac Yeffet, of the El Al Airline and a former intelligence director with the Israeli Secret Service, Yeffet openly questioned the unjustified focus on technological fixes. As he put it, no system in the world can replace the experience and judgment of a human agent. It can be argued that the huge sums of money and tactics such as removing your shoes for the security screener, is just window dressing to pacify the public.

North Americans have it relatively easy compared to passengers flying on El Al and the proverbial unfriendly skies are no match for the security you would encounter flying in the Middle East. Far from a 30 second cursory check, passengers going through the screening process to board an El Al flight can take anywhere from ten minutes and often much longer. And the success of their security is stuff of legend—there hasn't been a bombing on an El Al flight since 1968. For example the reliance on human intelligence would have alerted agents that the hijackers were traveling first class but did not look wealtyh enough to pay the fare and red flags would have been raised. I doubt the hijackers would have even got into the airport let alone the aircraft. Pan Am, ironically, rejected a security plan drafted on the Israeli model in 1987, one year before the Pan Am bombing over Lockerbie, Scotland that killed 270 people.

I am not sure, in such a politically correct climate, whether North America could stomach the medicine offered by the Israeli's; Americans might be inclined to balk at El Al's openly discriminatory practices and ethnic profiling; passengers on El Al are classified into three categories: low risk for Israeli or foreign Jews, medium risk for non-Jewish foreigners, and naturally high-risk for anyone that looks or sounds Arabic. Those identified as high-risk are immediately taken into a room for a "robust" body and baggage search and a *very* lengthy interrogation.

The point of all this is very simple: CAPPS and many of the systems I've discussed so far do little to enhance the security of passengers but they do serve to collect an enormous amount of information. Whether or not that information has been useful will never be known—and in fairness to security officials, it is not as though they can publicly discuss how it functions lest they risk providing too much information.

We are not going to see tangible improvements in aviation security until we let go of the obsession with security technology and return to the fundamentals: militarily trained security staff with significant years of experience to replace the legions of underpaid, and inexperienced screeners that we have now. Computers are only as good as the information they hold, and if someone wants to assume an identity then they will—only a well trained agent could pick up on the anomalies in a person's story or their facial features. Computer systems shouldn't be abandoned but we have to let go of the fallacious attitude that they can replace staff.

There is a good reason that El Al serves as the aspiring model of any airline—it works. That might seem like a contradiction coming from a privacy advocate but I would be less offended by a 15-minute interview with an agent than I would the myriad systems that are nosing into matters not related to my flight. Systems that check if I have parking tickets or owe back taxes while I board do nothing to improve safety in the skiers. The only way to truly improve aviation security in the United States is recruitment and a new philosophy towards protecting the airways.

6.4 TIA: Total Information Awareness

"There is much pleasure to be gained from useless knowledge."

—Bertrand Russell (1872–1970)

There are two more systems discussed in this chapter that are so large that they defy description—systems that have the potential, if ever fully unleashed, to track virtually every single detail of a person's life; where they went to school, comments from teachers, grades, work histories, comments from the boss, where they drink, who they associated with, books, movies and so on and so on. The system in question, however, is not run by the Department of Justice or even Homeland Security—it is run by the Department of Defense.

The driving vision of the *Total Information Awareness* (TIA) project was to provide the United States with the ability *to detect, identify, classify, and prevent terrorist activities before they occur* and is based on the assumption that even terrorists must leave a substantial *data trail* in the course of their planning—a data trail that researchers believe that can be detected amongst the trillions of transactions that are conducted daily. In order to discover these patterns the researchers at the Defense Advanced Research Project Agency (DARPA) envisioned an ultra-large scale database environment that would track every transaction by every citizen in the country. And somewhere in that massive sea of data would be clues such as the purchase of airline tickets, rental agreements and credit card transactions. It becomes all the more troubling when you discover not only how it works but also who runs it.

As it was mentioned in Chapter Three, the **Defense Advanced Research Project Agency (DARPA)** is the Research & Development division of the Department of Defense and they are generally known for developing technologies that provide the U.S. with defensive and offensive advantages on the battlefield. They have a multitude of research groups of which the Information Exploitation group is just one—I wonder if they have a group that researches what to call these groups.

The problem is not with DARPA, they are a well—respected and exceptionally brilliant group of scientists—it is the *political* appointments and management approach that has raised a lot of questions. The underlying problem rests in the murky world where intelligence and politics comes together and the Program Manager they appointed to lead this initiative was none other than Rear Admiral Dr. Poindexter, the most senior official of the Reagan administration to be convicted on five felony counts of lying to congress, destroying evidence, and obstructing congressional inquiries into the Iran-Contra scandal. Even Dr. Poindexter's appointment was highly suspect; while official documentation shows that he was appointed by Dr. Tether as a Section 1101 Appointee under 5 U.S.C. of the Strom Thurmond Defense Appropriation Act of 1999, the Tech Law Journal points out that Dr. Poindexter approached DARPA while working for a defense contractor under DARPA. The water seems to get murkier and muddier by the day. Most convicts would be hard pressed to get a job at the drive through window at McDonalds, so how did a convicted felon land a plum assignment to head up the largest surveillance network in American history. As a frequent visitor to the DARPA site, it look little work to discover an unusual "special program" called the "Experimental Personnel Hiring Authority" that falls under Section 1101 and allows for the *streamlined* hiring of "eminent scientists". It seems more like a neat way to sneak your friends into the backdoor of the movie theatre.

It has been argued that focusing too much on Dr. Poindexter is playing into the fears of the public and perhaps that criticism has merit; I should stress for the record Dr. Poindexter's convictions were ironically overturned on constitutional grounds. Nevertheless, you have one of the most controversial and invasive systems ever built with the theoretical ability to collect the most sensitive information that we have, led by a man who has a history of lying and selling weapons to third world countries—a dark past at best. But if challenging the leader seems unfair then I will focus on challenging the vision and the technology behind TIA—that is certainly more objective. According to Dr. Poindexter, information technology is the answer to terrorism and that the technology has the potential to solve many of these problems related to terrorism.

This belief that technology solves everything is the ranting of the technophiles and TIA is a perfect case of technology run amok. Likewise, the complexities of the social and political issues become lost in lines of code; terrorism, or any other security threat, cannot be solved by technology alone—something the CIA has learned the hard way. These problems are best attacked through a fusion of foreign (or domestic) policy, law, and social change. In the case of terrorism, no satellite, no matter how sophisticated, can identify a target unless the target has first been positively identified by human assets on the *ground*. Indeed one of the reasons we have not faired as well in the Middle East is the focus on electronic rather than human assets. It does imply there is no place for TIA; actually I think TIA is an indispensable tool—so long as it is directed outwards and combined with an increase in human assets in trouble spots around the world.

But the technology has raised its share of questions in the scientific community because TIA relies on several technologies that haven't even been discovered yet—TIA is a collection of disparate initiatives aimed at revolutionizing advances in computing, information collection, and analysis that was envisioned to give government the ability to sift through the affairs of the population to weed out threats. The TIA Program is comprised of several systems including databases and data mining environments that include data referred to as "all-source information", which refers to every type of information ranging from security records to images, phone conversations, and video clips; it would contain everything, or have the ability to acquire everything, from medical records and prescriptions to school grades and records of library books. Based on DARPA BAA 02-08 (Broad Agency Announcement) the project was primarily focused on three key areas including *repository technologies* (databases), *collaboration and automation tools,* and *Prototypical Closed Loop Systems* (finished prototypes). Of the 180 project submissions, 26 were approved for funding.

It was the Association for Computing Machinery, one of the most venerable organizations in the field that comprises the best minds in computer science that made the most valid technical observations. The position of the ACM held that

TIA suffers from serious "fundamental structural problems". It even noted serious doubts whether TIA can achieve its objectives in addition to the vulnerabilities that TIA may itself expose—in the end the ACM questioned whether TIA will cause more harm than good and their position is about as objective as one will ever find in Washington.

There are, as to be expected, economic and legal issues that have not been adequately addressed. The perception of excessive surveillance and the knowledge that every movement is watched runs the risk of altering the collective psyche of Americans—and a skittish population will wreak havoc on a market that relies on consumer confidence. The U.S. further runs the risks of alienating international trading partners and the European Union's Data Privacy Directive could be a cold bucket of water on American surveillance aspirations.

To get your mind around the immensity of TIA, let me summarize the program differently. In essence you have the U.S. *military* building a massive surveillance network with the express purpose of tracking *Americans*. You have military research agencies like the U.S. Navy Space & Naval Warfare Systems Laboratory (SPAWAR) involved in the program, either past or present, not to mention other interesting relationships. It is worth noting that, historically, the United States military was once forbidden from acting on domestic soil under the terms of the Posse Commitatus Act of 1878 because of abuses that came to light during the Civil War. But since the 1980's, in concert with the "War on Drugs", the military has increasingly played a larger role in law enforcement on domestic soil and there is a long list of incidents including several deaths that have been attributed to this encroachment. There are murky references to national security and clear and present threats that would allow military forces to move into urban areas if the political will existed. TIA is really an extension of a philosophical shift in domestic policy and all of this has been going on long before the threat of terrorism dominated the security stage. There is always going to be, it would seem, some war—first it was the "War on Drugs", then the "War on Terrorism", and who knows what will come next.

In speaking of the connection to the United States military, I run the risk of upsetting a lot of people and terms such as unpatriotic will inevitably make their way to the surface; so it is appropriate to launch my own pre-emptive strike by emphasizing I am actually quite a *Hawk* and I openly support U.S. military action when there is a *clear* threat to the security of the United States. But that raises the question of what constitutes a threat: in my opinion Iraq never was a threat to the U.S.—and the lack of chemical or biological weapons would seem to reinforce that position. Nation states like North Korea or Iran represent a much greater risk to America than Iraq ever did. Organized crime, drug syndicates, and terrorist cells also present a clear danger to the security of Americans and I would expect U.S. forces to eliminate those threats. But to turn the muscle of the U.S. military on its own population is not the answer.

So where then is TIA today? Eventually the Senate and Congress ordered a halt to TIA and suspended funding of the program *in its current form*. But TIA is by no means dead and is likely to resurface under a new name, or a new agency. There is also the possibility that TIA could go "dark" and disappear into the labyrinth of black programs. Too much time, money, and political capital has been invested in TIA to see it destroyed so easily and I would fully expect the program to re-emerge at some point.

6.5 Echelon

"The superior man, when resting in safety, does not forget that danger may come. When in a state of security he does not forget the possibility of ruin. When all is orderly, he does not forget that disorder may come. Thus his person is not endangered, and his States and all their clans are preserved."

—Confucius (551 BC–479 BC)

No discussion on state surveillance and privacy would be entirely complete without a brief discourse on the surveillance network that doesn't really exist—well it does but until recently its existence was vigorously denied. This is a network so vast that it spans continents and countries, surface and space; it has the capability to monitor every conceivable type of communications signal ranging from cell-phones to the Internet, faxes, and telexes. There is no oversight, no rules, no senate committees, and seemingly no constraints that govern its use—of course how can there be, it doesn't exist. Until recently every participant of the network had denied that there was ever such a ridiculous idea to begin with.

The system in question is commonly referred to as Echelon and is used as an automated global interception and relay communications network that is managed predominantly by the U.S. National Security Agency and partnered with five nations including the Canadian Security Establishment (CSE) in Canada, the Communications Headquarters (GCHQ) in Britain, the Australian Defense Signals Directorate (DSD) of Australia, and the security services in New Zealand.

There certainly could be other countries involved in the network that are signatories to several international agreements. Its origins appear to have had their beginnings in the UKUSA Agreement of 1947 but Echelon itself was not created until 1971. By 2004, it is generally accepted that there are at least 30 other nations with their own communications intelligence networks that at the very least have the ability to collaborate or provide information feeds to Echelon.

It isn't really that difficult to confirm *fragments* of this shadowy world—you can't hide a multi-billion dollar network forever, though it does require sifting and connecting through endless government documentation; even then there is very little to tie strands of evidence together in a manner that would provide a complete and compelling story. Even as I discovered, there is enough anecdotal proof to make the supposition that it exists but not enough to pull it together. For example, you can't just "ask" the government for technical specifications on this network, it is unlikely they will be very responsive or appreciative. But a more indirect route would reveal a tracking station in Sugar Grove, West Virginia…and the only way to know that would be to know: a) that there was an installation, b) that it is maintained and operated by the Naval Security Group Activity (NAVSECGRUACT), and most importantly c) much of their mission and functions can be found in NAVSECGRU INSTRUCTION C5450.48A—their training manual. To truly gain a sense of the magnitude of this network in the United States alone would mean identifying *every* military communications and intelligence "entity", assuming you could, and then determining if their mission and functions had anything to do with Echelon—even then, it would be the tip of the iceberg.

At a technical level the network, at least based on publicly availably accounts, is beyond fascinating and it would be difficult not to appreciate its complexity. We are talking about massive ground antennae that capture satellite transmissions around the world coupled with over 120 space-based satellites that capture global ground communications—not to mention the alleged taps into surface and transoceanic trunk lines. It can capture, according to speculative sources, 95% of the Internet's traffic and three billion communications messages a day, relying in large part on artificial intelligence and MEMEX pattern recognition algorithms to filter the traffic and find messages of interest to specific countries. When speaking of A.I., I am not referring to some omnipresent artificial species—having some relative experience in this field, I am more inclined to believe that they are using a combination of neural networks to classify traffic and evolutionary programming like genetic algorithms to refine the rules.

Suffice to say, this is just the tip of the iceberg. The French, who have always been quite vocal about Echelon and have not shared in the United States penchant for secrecy, are no slouches in this regard and have developed their own network with 12 known bases and the Helios 1A and 1B satellites and the Karou space station. From this it would not be a stretch to speculate that every major, developed country has own their capabilities for monitoring traffic.

The question then arises, why haven't you heard of Echelon or those others? First of all, it doesn't officially exist and it remains highly classified—even if the walls are starting to come down. Already the Dutch and French have openly admitted, much to the probable annoyance of the U.S., the existence of the network and both Australia and New Zealand have also divulged that it is indeed quite real—only the U.S. continues to insist that it's a figment of everyone's imagination. I would safely suspect that there are many in the intelligence community losing sleep—because the revelations have resulted in a renewed interest from congressional and senate authorities and privacy groups and that can't make the NSA very happy.

While there is little that is available, there is enough documented evidence to indicate that all of the participating countries share "dictionaries" of key words, people to be tapped, and similar requests which are forwarded to the requesting countries agencies. This makes moot the argument and laws that intelligence services cannot conduct surveillance on their own population since they don't have to; they can technically request their allies to do it for them. The requesting country can also specify what type of communications they want to receive—for example a Telex is called a MayFly and a phone tap is called a Mantis. It all makes for really interesting cloak and dagger stuff.

Even though the network was and likely still is focused on military intelligence, there have been a lot of accusations that the surveillance network is also being used to gather international economic intelligence—a claim the French have made on many occasions, even if they probably do precisely the same thing.

The United Nations has also investigated claims that Echelon was used to spy on members negotiating the GATT treaties.

That is one of the key problems with Echelon—if ever there was a surveillance mission capable of "mission creep", Echelon is it. What began ostensibly as a military intelligence system could easily be expanded in scope to collect the communications of anyone it wishes. Since the network is so classified we may never really know what rules there are or whether Echelon is constrained in any way against abuse. I am inclined to believe that the capabilities are directed at more important intelligence than all the junk mail in my inbox or my phone calls to have my truck fixed. Then again it would seem they could do so if they see fit.

The greatest difficulty with Echelon is that lives in a dark, black world devoid of public access, scrutiny, and oversight. It is so secret that, despite the fact it has become known, the National Security Agency will still not acknowledge that it even exists let alone what it does. Even releases of government documents leave much to be desired when those documents are so censored that you may as well be reading a black marker. There have been some releases under the FOIA legislation and more detail has become known under the Scientific and Technical Options Assessment reports but it still leaves us very much in the dark. As the system grows, as it has with the addition of OASIS and FLUENT, it opens the question of mission creep. A system originally designed to spy on the communications of suspected persons could easily be expanded to simply spy on everyone as it has been accused of doing.

With little or no oversight outside the intelligence community, it opens the possibility of spying for the sake of spying. There have already been a small trickle of press reports (which never seem to reach CNN and other major networks) including congressional testimony that have accused the Echelon network of spying on U.N. diplomats and possibly engaging in industrial espionage. How hard would it be to target trade negotiators rather than terrorists? How hard would it be to spy on anyone for that matter? It would seem there is little this network cannot see or hear.

How do you even defend yourself against something like Echelon? Technically you could encrypt your e-mail (as I often do) or even buy scrambled phones—with the understanding that the NSA can crack a 56-bit key in about three hours if not less. You could go with a "hard" key over 1024 bits but the problem is this is not knowledge generally held in the public domain. I am pretty sure my mother has never heard of a crypto system let alone how to set the key on a STU-III phone. The fact is, there are just some things we can't control, and this is probably one of them until there are meaningful investigations and debates on how intelligence is gathered and used. And there is the other side of the coin—how much control do we want to have? Do we want to limit the abilities of our intelligence agencies in collecting intelligence in the interest of national security? I would not seek to inhibit or limit the effectiveness of the intelligence community, I would just prefer they leave people alone that clearly do not constitute a threat.

6.6 End Game

> *"The trouble with fighting for human freedom is that one spends most of one's time defending scoundrels. For it is against scoundrels that oppressive laws are first aimed, and oppression must be stopped at the beginning if it is to be stopped at all."*
>
> **—H. L. Mencken (1880–1956)**

Well, perhaps my title was premature, there is no end game; this is an eternal game of cat and mouse that will continue for decades if not centuries. And it is a dangerous skirmish between governments and insurgents, regardless of type or cause, and the population is frequently caught in the crossfire.

One of the greatest challenges facing any democratic government is the balance between the expectation of freedoms that are enshrined in law and the need to provide for the security and safety of its citizens. For the average person, the only exposure they usually have to law enforcement is either parking tickets or a fast read of a shootout in someone else's neighborhood—it's always someone else of course. But there are those that mingle amongst us who do not share our view of the world and their aim is to disrupt and destroy our way of life. This is not a new battle and the challenge faced by counterintelligence services has been to find those that would cause harm while being shackled by the protections afforded by the Constitution. So the cat and mouse game continued unabated right up until September 11th, 2001. And then the proverbial cat was given some new toys so to speak.

As is so often the case, a pendulum must swing to both extremes before finding a sense of balance and the same can be said of security. At one end of the spectrum there is a utopian view of a state in which freedom and privacy are absolutes and at the other end of the spectrum there is the proverbial "police state", Orwell's 1984. Since 9/11, democratic nations have clearly swung towards a police state and there is little evidence to suggest that things will return to normal any time soon. The sheer momentum of events coupled with the philosophical direction of the current U.S. administration ensures this pathway will be followed for some time, or least until the next election. Even if President Bush was defeated, the infrastructure is in place and the mindset has been entrenched for decades.

Perhaps surprisingly, many of the rights enshrined in the Constitution's of Canada, the United States, and Britain, are not absolute. On the contrary, each country has provisions and exceptions that clearly outline conditions under which security and safety takes precedence over privacy and civil liberties. In the United States in particular, the passing of the Patriot Act gave the government unparalleled powers of surveillance and arrest in addition to the ability to detain suspects indefinitely without charges or access to attorneys in some cases.

Likewise, there has also been a renewed attention given to the use of advanced technologies to aid in the surveillance of email, phone calls, and other forms of communication for the presumed purpose of identifying threats. There are other systems that track our flights and movements while others still can listen to cell phone conversations. Ultimately such changes to our freedoms mark a compromise of our liberties in exchange for a sense of perceived security; whether we are any safer is open for debate and it is unfortunately a debate that is not taking place except between the government and organizations like the ACLU, and the few of us who read their position papers—it hasn't, with the exception of the occasional CNN headline or opinion, reached the mainstream media.

During the course of this chapter, you have been exposed to a few of the principal systems that were designed to provide government with an overreaching view of every citizen in the country, even when these endeavors do not exactly work as they were supposed to. I have been tinged by a little bit of guilt for focusing on the systems rather than the legal and social impact of surveillance but I felt it would be best to leave those topics to those better educated in such matters.

From a technical perspective, in some cases such as TIA, these systems are, at best, overly ambitious and far too complex and risky to really take seriously. In addition to the technical challenges of getting it to work, some of which don't even exist yet, there are also problems associated with project management, planning scope, budgeting, and of course the mind boggling investment in money, materials, and human resources required to build and deploy these monstrous initiatives. There is also the threat that these systems could be exploited for the purposes of identity theft and other crimes by internal employees and external forces. Even if the technical, security, and logistical issues could be resolved there is still myriad ethical and constitutional obstacles that must be addressed.

The fact that these initiatives are raising so many questions is actually a good sign that there are still those that are working to keep a lid on the scope and reach of systems that seem to lack a conscience. Actually it is the sponsors that lack a conscience—and this is exemplified by the rationales they put forth—in their world anything that improves security is a good thing, regardless of the social impact it might have. And it is for that reason court actions have been on the rise to seek injunctions against these projects and bring them to a halt   at least long enough to objectively assess their future impact. We have already seen California place a moratorium on funding for public biometric systems and Congress put a temporary halt to the development of TIA.

Unfortunately, there seems to be pattern developing in which the developing and sponsoring agencies build the systems, leak the news, wait for the protests, and then ride out the "storm"—and then wait for the publics attention to wander elsewhere—and the media fuels this lack of attention by not fully investigating and continually reporting on major stories. It is this lack of objective deep journalism that allows these scandals to eventually fall off our collective radar only to be reborn under a new, more obscure name that may never see the light of day. Of course it would do the public well if these agencies developed a concept, released it to the public, sought public consultation, and built in the necessary privacy protections during the design phase rather than after the fact. Maybe I am being a tad too cynical. Whatever the case, it is for the reasons I've citied that I believe we require an aggressive press and an even more aggressive network of privately funded third party organizations like the ACLU to bring these programs to the press and educate the public. That is how TIA ultimately met its (quasi) downfall. If was not for these groups, programs like TIA would have gone on without us never knowing any better.

Think of the social consequences and costs that stem from the genesis of a surveillance state run amok. Firstly the slow incremental nature of progress means that many of the changes that occurring remain invisible until its too late to turn back the clock—and society finds itself stripped of one less liberty that was once considered sacrosanct. Over time you begin to see more and more cameras in the malls, at the park, and on the street corners—and eventually this extends to every block, then every building, then the rooms within the building—and you can see where that is going. Every camera, every microphone, and data file results in decisions and actions by people we will never meet and which affect your quality of life. Those actions are certainly not benign; I am referring to unpleasant actions like appropriation, confiscation, humiliation, intimidation, and ultimately the loss of liberty. Asset forfeiture has become rather profitable for police forces through the United States because, put simply, the police can seize virtually anything they wish based on the extremely low legal standards of probable cause—boats, cars, homes, and money.

Beyond the physical forfeitures, what effect do such programs have on the collective psyche of a country—will you move as freely or speak as freely as you once did? Will you become more guarded in your choice of words knowing your phone conversations could be tapped? This is a form of behavioral modification—it isn't new, it is just being reinforced by new technologies.

The end game, or at least this chapter of it, is a country and society that none of us approved of or voted for; a society that seeks to constrain and control rather than liberate. Is it not ironic that North America, which claims, often loudly, to be the most democratic in the world, possesses the ability to be the most watched in the world? That is not to say that other countries are not worse—you need only look at Iran or China to understand what a controlled society really looks like that ought to engender appreciation for the freedoms we do have. The objective is to ensure that the liberties provided by the constitution including the expectation of privacy are not diminished or eroded by those that do not see quite the same value and importance in the constitution that is shared by the majority of the population.

Protecting civil liberties and privacy is not a part-time position and it requires the courage to challenge authority if that authority is no longer acting in the interests of the public when it was chartered to serve and protect. Let there be no confusion in my choice of words—for all of the lofty systems and ideals, what remains at stake is the right to travel freely without having to surrender constitutionally guaranteed protections. You do not have to understand the intricacies of data warehouses to appreciate your right to get on a plane to fly from Toronto to Vancouver or from New York to San Diego without having your every movement tracked if not prevented entirely—CAPPS could achieve both without anyone realizing it before its too late to turn back.

7

Best Practices and
Recommendations

7.1 A Better Way

"There is no law of progress. Our future is in our own hands, to make or to mar. It will be an uphill fight to the end, and would we have it otherwise? Let no one suppose that evolution will ever exempt us from struggles. 'You forget,' said the Devil, with a chuckle, 'that I have been evolving too."

—William Ralph Inge (1860–1954)

Granted, a majority of the chapters that I have covered so far have focused upon many of the philosophies, paradoxes, and abstractions of privacy; they are not necessarily practical but they are crucial to understanding this phenomenon. Chapter Seven marks the final stretch and offers a combination of information ranging from the practices of workplace surveillance and privacy and the threat of identity theft to a series of best practices that have been collected from a variety of authoritative sources and which present the best known techniques for protecting privacy. Though not quite as "exciting" as smart cards and spy satellites, the best practices provide the most practical measures that *everyone* reading this book can adopt as their own.

As to whether a person or company elects to put these changes into effect or take seriously the recommendations put forth is entirely a matter of choice though more so for the former than the latter. For the individual, it is critical to recognize that the first line of defense rests not with statutory restrictions but the level of vigilance enforced in all manner of professional and personal affairs. I, nor anyone, else can force or compel readers to take responsibility for their privacy; it is up you what you do with it.

For organizations, the challenge is no less daunting. On the one hand, firms are required by law, dependent on location, to exercise great care in the collection, use, and disclosure of the personal information that, in the end, *belongs* to their customers or staff; violating that trust can result in a multitude of consequences ranging from government investigations and financial penalties to litigation or possibly both. Recognizing this concept of ownership is, in itself, quite a leap for many in the private sector. And yet every commercial entity also faces fierce competition that demands business intelligence to effectively compete in the market, which in turn results in suffocating constraints put upon a company. Then again, in light of Enron, Tyco, and the endless spectacle of corporate scandals that have erupted in the past few years, consumers are already wary of the private sector as it is and it is incumbent on the business community to rebuild its trust with the public.

Identity theft, given its catastrophic impact and its relationship to privacy is the first topic on the deck and it is an ugly one at best; it is very difficult to prevent, and especially difficult to correct after the damage has been done. This phenomenon has become an epidemic and the risk of doing nothing could invariably cost you your job and everything you've ever strived to acquire. In the strongest terms possible, identity theft has become a pandemic costing billions of dollars world wide, and leaving in its wake many people whose lives have been devastated by financial ruin.

The trend, emerging in the 1990's, has been constantly growing in the past decade as identity thieves have become more sophisticated in their approaches and techniques and it has taken the law some time to play "catch up" to the point where it is now a serious federal offense carrying hefty prison terms and fines. As it turns out, there are various defensive strategies you can implement to protect yourself accordingly and minimize your target profile so that they might move on to a more attractive target.

As for the abyss of workplace privacy, it is undeniably a touchy subject that has a habit of polarizing opinions on either side of the management divide and in the course of tackling such a thorny issue it has proven to be more of a challenge than was originally anticipated. The polarization of positions is a natural occurrence—a manager will never be able to see things in the same manner that an employee does and vice versa, so depending on your role, this segment will either sit well with you or it won't but it doesn't detract me from the need to cover it. Unfortunately, violations do indeed occur and if they didn't there would be little need to wade into this battle—the sheer volume of lawsuits and complaints are a testament to a growing problem and must be addressed.

In all fairness to the private sector, or any organization for that matter, there must be agreement on why a business exists in the first place and that is, first and foremost, to make the owners wealthy. They invested their money, they took astounding risks to achieve what they have created, and it must be understood that company's property was acquired for the express purpose of achieving that mission. Lest anyone forget the object of the game is wealth creation and obstacles interfering with that objective will surely be confronted and eliminated.

Somewhere along the continuum of opinions, there is a boundary between appropriate measures taken by a business to protect its assets and inappropriate intelligence gathering measures that serve no legitimate purpose except to discover the thoughts and habits of their employees. I will leave the rest of my thoughts on that until we come to that discussion.

In terms of the best practices discussed in this chapter, I offer two diametrically opposite perspectives and related recommendations, one set for consumers and the other set for business, with the expectation, or at least the hope, that in both we can find a reasonable compromise between the need for survival and the need for privacy. It is not my expectation that everyone will agree with every recommendation that has been made but these are, nonetheless, commonly accepted guidelines provided by virtually all privacy advocacy groups and privacy commissions.

As for the best practices offered to consumers, they are there to use as you wish—every one of them is presented to limit your exposure and bestow a peace of mind as much as can be achieved by a book. Whatever my thoughts on the matter, no one can conduct themselves in society without sharing information, that wouldn't be reasonable or realistic. But there are things that can be done to mitigate how much light is shed upon a person's most private domains and preserve your privacy. Having shed light upon many of the challenges to privacy in previous chapters, it wouldn't be surprising to discover many readers are disheartened and resigned to the destruction of their privacy rights. Albeit there are serious threats against privacy but it does not necessitate the abandonment of core ethical values nor does it mean that you are bereft of defenses.

The guidelines offered to organizations are based on a different set of concerns that focus more on the trend towards increased legislation and regulatory control based on many of the principles discussed. Transparency, security, scope, and consent are legally mandated in many acts in Canada, the United States, and the European Union. If that is the proverbial stick, then the potential for improving customer retention and revenues through proactive privacy policy management is the carrot. Businesses, unlike their consumer counterparts, may soon have little choice as to the practices they adhere to.

If there was a key differentiator to be found between this and previous chapters it is clearly that this chapter focuses on things that can be controlled and changed and over which there is a degree of choice. There is little that can be done to stop governments from putting spy satellites into space or reading everyone's mail, anymore than we could prevent drilling in Alaska—some things are beyond the control of the individual. That does not mean that consumers, employees, and businesses cannot take control over many aspects relating to privacy and these guidelines, if only a starting point, offer the means to constructively change practices and policies. That is better than doing nothing.

Before getting into the heart of things, there are no promises made or implied in these recommendations. Remember that the idea of absolute security is a fallacy at best and leads to a false sense of security both for people and organizations alike; you can mitigate and reduce the probability of a threat but short of hiding under a rock, there is always going to be that risk. As for compliance with the respective laws of your region, best not to seek that from me, I have trouble remembering where I put my parking tickets, so for such matters an attorney is recommended. That said, lets begin with a look at identity theft.

7.2 Combating Identity Theft

> *"The particular human chain we're part of is central to our individual identity."*
>
> **—Elizabeth Stone**

For every moment that passes, someone, somewhere, becomes the unfortunate victim of identity theft, and in it's wake, it leaves many if not all in varying degrees of financial ruin—it can take months if not years to undo the damage caused by identity thieves and it is incumbent upon you to safeguard your privacy to prevent it from occurring in the first place. My objective at this point of the game is to hopefully impress upon you the dangers of identity theft and offer a series of countermeasures to prevent it from happening at all.

Identity theft is, put simply, defined as the collection, use, and dissemination of personal information of another person for criminal purposes but this definition fails to illustrate the epidemic nature of the problem and the dire consequences it has for individual victims, organizations, and society as a whole. And when speaking of personal information, as I have described in earlier sections, I am referring to information that isn't privy to the public—your name and address are really the only publicly available pieces of information—everything else is private; and that includes everything from demographic data (age, sex, marital status) to your credit cards and other sensitive documents.

Despite the presumably private nature of this information, it is quite remarkable how easy it is to gather enough personal information about a person to begin the process of creating a new phantom persona. Once a thief has your name, diligent research can uncover whatever is necessary to request new birth certificates, and this in turn provides the gateway to all the other critical forms of identification they would need to conduct illegal activities in your name. Everything we do tends to leave an exhaustive transaction trail, and those in the field are quite adept at discovering who you are, where you were born, your maiden name and whatever else that is required to begin the process of stealing your identity. It is this collective patchwork of fragmented data elements that is immensely valuable, and there are measures that can be taken to prevent unwanted disclosures and thefts.

It cannot be stressed enough how sophisticated these groups are; in fact, it is important to grasp that this is not a case of amateur criminals but organized crime syndicates with the resources and manpower to conduct significant research activities and surveillance to achieve their objectives. The field is sufficiently advanced to employ "pre-cursor" organizations that collect and compile "baseline data" like where you live, your age, and other facts, which are then sold to *secondary* groups that use the data to duplicate your identity. Those identities can then be sold to other criminals or used to perform a broad range of crimes ranging from evading capture to perpetrating fraud.

The good news is that law enforcement does take this seriously and identity thieves can be prosecuted under multiple federal and state laws including the Identity Theft & Assumption Deterrence Act of 1998 which carries a maximum term of 15 years in addition to fines coupled with the likelihood of charges not limited to theft and wire fraud depending on how the identity was captured or used. The seriousness of the crime is particularly well illustrated by the combined efforts of the FBI, Secret Service, Treasury Department, U.S. Postal Service, and other agencies that work together to solve these offences.

There is a lot they could do to prevent these things from occurring in the first place and much of that effort should focus on the re-engineering of source documents like your birth certificate, drivers license, and passports—as it stands now, in many states and provinces, it only takes a single form and knowledge of a person to get a birth certificate—and from there it is a small step to social security cards and credit cards. On the bright side there is a lot of movement towards this and I am aware of several initiatives that will eventually bring about biometric passports within a few years; if they are designed properly it will make the forging of passports all but a footnote of history.

So that I can eventually end this segment on a positive note it is time to focus on what you ought to be doing in the event you suspect you have become of the victim of identity theft and the steps you need to take to regain your name. More importantly, provided shortly will be a series of defenses and countermeasures to prevent the unthinkable before it happens.

The first and most important step, one that is tied to protecting yourself is absolute vigilance. It is imperative that you inspect every bill you receive to ensure there are no "anomalies"—are there unusual purchases on your credit card, have you received unusual phone calls from creditors you have never heard of, are you receiving confirmations for address changes that weren't of your doing? You must be mindful of changes that shouldn't be happening and which are often the first indication that there is a problem. Identity thieves are a resourceful, if not scurrilous bunch, that are quite capable of ferreting any type of information. To name just a few of them:

- The theft of mail is the easiest and provides a wealth of very sensitive information included bills, invoices, and whatever else you throw out. Do whatever is necessary to protect your mail and check with vendors, merchants, and the post office to ensure your mail is not being redirected without your permission.

- Euphemistically referred to as "Dumpster Diving", thieves will routinely search through the garbage for discarded documents that offer clues to your identity—what have you thrown out of late? If that question is answered as I suspect it will be, consider a personal shredder.

- Bolder tactics include stealing information from employers, which are made all the easier with customer databases and information repositories. In some cases, employees can be bribed to provide access.

- There have been documented cases of employees and their cohorts retrieving credit reports to identify potential victims. Several cases involving employees working at credit bureaus have led to convictions already and, while said bureaus generally strive to protect their data, an insider is tough to beat.

- They may engage in "phishing" or similar scams that may include posing as a legitimate organization to gain information. There is all sorts of phishing expeditions abound but the most common are the e-mails from E-bay or well-known bank (or whomever) claiming your account has to be verified and that personal information is required to confirm it. Don't even think about it, either delete it or take the time to contact the real company they are pretending to represent.

If you have determined or suspect you have been the target of identity theft, reporting and documentation are your most formidable weapons—keep in mind that creditors, the police, and government agencies, only have your word against a documented transaction trail—the more concrete evidence you have the better the outcome. You must notify the police immediately and document every step you take to stem the damage, including your expenses, which can be exceptionally high.

You will also have to take the painful step of contacting your bank and other agencies and that may include the painful task of canceling virtually everything including credit cards, debit cards, passports—every piece of documentation you have is likely compromised at that stage—keep the documents for evidence obviously but immediately begin the process of completing the necessary paperwork. You must also notify all of your creditors including phone companies, cable companies—everyone you do business with and then open new accounts so you can continue to function. Even completing all of these steps will not restore your name—and there are countless incidents in which people have taken months, sometimes years, to undo all the damage that has been caused. To make things even worse, the authorities may not be quite as receptive as you would expect—the evidence, the transaction trail, says you made purchases, and you are claiming you didn't—you can already see where that is going to lead. The evidence you gather and the steps you take will provide you with at least some defense against irreparable harm and do much to provide police and other agencies with the assurance that the case is legitimate. Given all of the headaches and heartaches that can result from these thefts, hopefully you will focus even more on defensive measures to avoid these problems from happening to you.

The ultimate defense to identity theft is minimizing your risk by minimizing your exposure—and that means revealing facts about you on a strictly need-to-know basis and only to trusted parties—sparingly. I am not referring to friends and family but vendors and unknown organizations or people that you have never dealt with before—if they aren't in your trusted circle than they are, by inference, un-trusted. When you are asked for personal information, you must learn to ask why it is required, and if push comes to shove, question the legality of the request. How is your information to be used and how is it stored? What third parties have access to transactional data? And whenever possible either refuse entirely or reduce significantly what you provide—a retailer does not need to know how many children you have, how old you are, how much you make, or anything else except your ability to pay. The same can be said for surveys, questionnaires—avoid them at all costs if they are not anonymous questionnaires because you cannot be assured where the data is going to end up.

Though much of this advice may seem like common sense there are still many, many people that routinely give out private information without questioning its need or the legitimacy of the questioner—you should always be questioning why they need to know where you live or how much you make; Simply walk away—disregarding protests to the contrary, and barring a few notable exceptions to the rule, I am always of the belief there is *always* the *same* product being offered somewhere else…and probably at a better price. Likewise, never give out your credit card number unless you absolutely have to, and never, ever over the phone unless it is a trusted party that you know and which whom *you initiated the transaction*. If the request is legitimate then call the person back but verify the number is truly that of the organization.

Of the most questionable practices, it should raise red flags when a merchant requests your social security or social insurance numbers and you need to understand that in most states and provinces it is illegal to ask for either that or your health card—unless you are seeing your doctor or visiting a hospital, never use your health card as a form of identification, and if you are a vendor, you should never accept it as a form of identification. Employers may also require identification or when you apply for credit—beyond that it is best to ask a lot of questions. It is not unusual for organizations to ask for private information in a manner that suggests authority or their information systems are constructed such that those fields must be completed. That does not require you to provide it.

There are also proactive defenses you can take to prevent further risk such as speaking to your credit card companies, creditors, to put traces in place, to call you when address changes are attempted, and to call if there is suspicious activity on your accounts. They should be offering this to begin with and many financial institutions have invested millions (if not more) in anti-fraud departments to call customers when strange patterns begin to emerge (like making a purchase at the gas station in New York and another purchase in Florida five minutes apart). You should ask if you can put passwords on your credit cards over a given limit and ask the same of your other creditors—it's a small price to pay for that peace of mind.

Entire books have been devoted to the subject and it is important enough to include here but I wouldn't want to get carried away, so let me spin it a different way and borrow an old but current mantra used at the NSA and elsewhere. One of the principle lessons that law enforcement and intelligence officers learn is the assumption that an opponent possesses infinite resources, manpower, and the time to attack you—you are taught to never underestimate your opponent and these are lessons that extend very well to everyone reading this book and privacy in general. Old but still true, an ounce of prevention is worth a pound of cure.

7.3 Privacy in the Workplace

"When a man tells you that he got rich through hard work, ask him: 'Whose?'"

—Don Marquis (1878–1937)

When it comes to privacy issues, workplace privacy is a potential land mine waiting to go off on the unsuspecting and it is just as controversial for authors that wade insanely into this abyss. Still, it would be indubitably negligent of me not to, so here it goes. Every journey has a beginning and the best place to begin is a definition of the relationship between an employee and employer regardless of the work being performed or how much is being paid. For the sake of argument, lets define employment as a form of contract between two parties, the employee and the employer—the former being contracted to complete a specific set of duties in exchange for an agreed upon set of compensation paid by the latter. Conspicuously, albeit intentionally absent, from this definition are any references to personal life of the employee for it bears little or nor relevance to the terms of the contract itself. This would be perfect if it was not for the annoying fact our lives are often a messy mix of triumphs and tribulations and that most of us, contrary to the thinking of a few, are not perfect automatons. For the most part it would be safe to say that a majority of employers generally meet their obligations, expect a reasonable days work of their employees—and more importantly, they would just as soon *not* know the lurid details of their staff and would prefer they act professionally and complete what has been assigned to them. But there are always exceptions to the rule, and if I come across as a tad harsh take comfort in the knowledge, especially if you are a manager, that whatever disdain I possess is focused exclusively on them.

Unfortunately, there are just so many complaints and lawsuits against employers for violations of privacy that it becomes quite clear that there exists this contingent of companies that possess an unhealthy animosity and disrespect for their employees which in turn leads some companies to discover, at every possible turn, every conceivable detail about their staff—including intimate details such as their lifestyles that have little or no bearing on the quality of work. So what could possibly provoke these companies and their management teams to engage in such offensive, divisive behaviors in the first place?

First of all, regardless of the rationales and reasoning, the weight of these decisions rests squarely on the culture of the company and that *always* permeates from the top—a function of corporate culture that is largely predicated on distrust. The reasons for such behavior are broad but arguments have been based on everything from improving productivity and operational efficiencies to protecting the integrity of a companies' public image. Ironically, surveillance strategies have the unpleasant habit of producing the opposite of the desired effect and can actually result in more problems than it seeks to solve, not the least of which is reduced efficiency, poor morale, poor productivity, increased absenteeism or lateness, retention problems, and so on. Indeed a strange way to shoot yourself in the foot.

At the more extreme end of the debate it has been argued that there are more "sinister" forces at work. The fierce competition of the marketplace has resulted in many questionable practices starting with "downsizing", "rightsizing", and "outsourcing"—all in an effort to create the leanest infrastructure possible and therefore allowing an organization to compete more effectively. Many of us, if not most, have already had some very personal experiences with these tired management "mantras"—those that are downsized find it difficult to find work in a tight employment market while those remaining must absorb even more work that ultimately results in increased stresses in the workplace. As one colleague put it so eloquently "My family is *not* an economic factor". But what does this have to do with privacy? More than you might realize. These very *same* competitive forces have resulted in efforts to eliminate "high maintenance" employees and maintain only those that we might call "super producers"—that ultra-efficient, never late, never sick, perfect employee. In other words, all of the surveillance and testing may actually serve to enforce a new form of "Economic Eugenics".

This entire idea of economic eugenics stems from the belief, in some circles, that there are segments of the population that are, to put it politely, *undesirable*. And the term *undesirable* is very subjective, depending on who is making the decisions. For the longest time society condoned and even encouraged overt discrimination against many sectors of society including minorities, women, and the disabled; that "umbrella" has actually gotten larger, not smaller. Included in that mix of people would be those that don't score well on the SAT, don't do well on specific performance tests, or worst of all, display a predisposition to genetic anomalies. Where society draws the line and places boundaries on what is permissible goes directly to the issue of privacy in the workplace; there must be details of the person that are off limits. That does not imply that employers do not have the right to protect their assets or hire the best possible candidate for a given position—it merely means that choices must not unduly discriminate based on a new set of criteria that would involve an invasion of an employee's privacy. And for that reason it demands a rigorous constitutional and regulatory framework to provide guidance for employers, consumers, and employees; if there were no abuses there would be no need for protective privacy and labor laws.

Such a small passage does not do adequate justice to the broad spectrum of methods and technologies currently in predominantly larger firms; the best I can hope to achieve is to provide a momentary snapshot in time—not enough to fully explain each but enough to provide a basic understanding of how privacy is affected. The technologies and tactics run across a continuum from the simple to the complex and from least to most intrusive—the most technological isn't always the most intrusive. Lets begin with the more traditional methods before exploring the more "glamorous" technologies.

The fact is companies need information to work effectively and that requires information from employees for taxation, interviews, resumes, exams, performance evaluations, and many traditional channels that form the majority of data that is required to function properly. However, in *some* environments this collection strategy goes far beyond what is needed and gets into some pretty murky areas and that is all the more critical when you consider that life altering decisions will be made and will be based on deeper background investigations that might include everything from criminal records and credit ratings to interviews with friends, family, and coworkers not to mention covert surveillance. In some instances, be it military, government, or intelligence positions, these measures are valid and necessary to ensure the *integrity* of the candidate. It is when these types of background checks cross into non-work related activities or for positions with no sensitive component that the darker side of human nature emerges yet again. As if that was not enough, there are some firms, including one automaker that shall remain nameless, that use investigators to determine the "lifestyle" of their employees—who are your friends? Where do you shop? Do you go to church? Who do you sleep with? Do you drink a lot? And the list continues ad infinitum. Although the latter forms of intrusiveness occur very infrequently, the fact remains they *do* occur, particularly at the largest of multinationals. The important point here is not only what is collected but what is done with the information—what decisions will be made based on these findings? Does your religion or lack of affect your career opportunities let alone job security—and ultimately *who* makes those decisions?

The next type of "intrusion" focuses on different types of "testing" in all of its forms including a variety of performance, aptitude, vocational, intelligence, personality, and integrity tests. In many cases this may very well be a legitimate requirement for entry into a position. A secretary must often demonstrate proficiency in specific administrative skills, bus drivers might need to go through a rigorous road test to ensure they are safe drivers—even I can find no fault with this type of testing if the purpose is limited to the evaluation of core competencies. It is *far* more questionable when employers start using other tests like intelligence testing (WAIS III) or similar psychometric testing methodologies—what scale will be used? What are the thresholds? If an employee scores an "average" IQ between 95-105 what's then? Is it relevant to their jobs? A mechanic might be exceptionally gifted and yet show an average of 97? How does an employer *interpret* these results? Likewise, the average IQ for a Ph.D. candidate is approximately 128 or higher—what if a candidate scores 125 but his dissertation is brilliant? As an employee it is important to ask the right questions—what will the results be used for, do you use qualified psychologists to interpret the results? What will be the impact on job security? For employers, the best defense, as always, is to communicate the reasoning for the testing, the standards being used, and what they really mean. Educating employees will save endless hours of questions and uncertainty.

When it comes to technological surveillance, it is safe to say there is absolutely nothing that employers cannot capture, trace, track, and analyze—especially when it comes to phones, faxes, computers, cell phones, and facility access. And in many cases, it must be made clear, these are *valid* means of protecting the assets of the company and make certain that employees are not misusing the very things that were provided to enhance the productivity and profitability of the company; the key is finding a balance. Employees must recognize they are there to work and that browsing the Internet for amusement is not considered a productive activity. Conversely employers must equally recognize that employees are *not* automatons—they have personal emergencies, medical conditions, the car needs fixing, the kids need daycare. Ironically, it is not as though employers don't have the same problems and needs, so a little bit of empathy and humanity will go along way to preserving morale and retention.

Phones in particular have always been somewhat contentious. Calls can be monitored for frequency, destination, duration, even which phones were used—and in many cases, it is possible that an employer is monitoring the *content* of the call itself. It must be noted that managers listening in on the calls of their staff are taking considerable risks—firstly, the interception of phone calls is generally illegal in both Canada and the United States except under specific conditions, most notably that both parties must consent to the interception. Secondly, it is equally possible that a manager could face not only charges under the criminal code but civil litigation as well—remember, just because an employee consented to the monitoring of personal calls in no way infers the *third party* consents to the monitoring—your legal headaches may end up with the spouse. For employees, there are certain safeguards you can take, especially if you know calls are monitored; most importantly minimize or eliminate personal calls at the office—better yet use a personal cell phone. Try to minimize the length of calls and keep the details to a minimum. And if you aren't sure about the company's policy either ask or simply assume they're listening—it will save a lot of headaches and possible embarrassments. As for employers, covert monitoring is just plain *illegal*—without exception, so now would be a good time to knock it off. Inform your staff about proper phone protocols and your expectations. If you do monitor calls for performance or customer service, provide a personal phone in a quiet area that is not, under any circumstances, monitored.

The monitoring of computers has exploded exponentially in the last 20 years, making it possible to capture everything from e-mails to keystrokes and everything in between. As it was with the phones, company computers were provided for company functions, not to e-mail friends all day, play on the Internet, or have live chats all morning. It is also more likely that managers or I.T. staff will have more intimate access to your computer than the phone so be careful what you store on them; it is no place for your intimate diary, "alternate" career opportunities, brushed up resumes, new game releases, pirated software, your favorite MP3 collections, and "other" types of entertainment. It never ceases to amaze me how many people are dismissed every year because they thought no one would notice the 20 *Gigabytes* of movies they stored on a server (as one defense scientist in Canada discovered). Likewise, where you browse on the Internet is easily and frequently tracked—safer still, just assumed it's logged. There is no need for generalizations on best practices except to say, know what is appropriate within your company. If your firm forbids personal browsing, just don't. If they do allow it, and they tend to be more relaxed, simply use common sense and don't abuse the privilege. The same applies to e-mail, perhaps even more so.

Facility access and control can be a challenging area, depending on the nature of the business and sensitivity of the products they produce—a pharmaceutical research group will require substantially more security than the producer of a box. Companies can employ a broad spectrum of facility access control systems that might include security patrols, smart cards, biometrics, pressure sensors, thermal sensors, numeric key pads, systems that track whereabouts and durations in specific locations, and list just goes on.

It would be difficult at best to criticize the right of a company to protect its physical infrastructure—and the use of the technologies listed above may be deployed to prevent access to controlled areas or notify security staff of breaches. It is far better to simply prevent access and explain why a person is not allowed into a given area than to play "cat and mouse". Where it becomes an issue is when the technologies are used to control behavior beyond what we would consider reasonable—tracking movement through a building could prove addictive for a micro-manager. How long did they spend in the bathroom, at lunch, in the hall. They are the same managers that would use "dummy cameras" and one-way mirror and the only advice I can offer is find another manager or another job.

While I have already discussed the proliferation of drug and genetic testing, it is so important that it deserves special mention here as a function of workplace privacy. First and foremost, no matter the type of testing, it is only valid if the testing is relevant to job functions where safety is paramount. Drug testing an accounting clerk is simply unacceptable and probably unconstitutional. Even if a person does test positive, it doesn't necessarily determine when the drug was taken and does not prove it was even recent. Are you to penalize someone for a spurious moment of youthful experimentation? What about the genetic testing of employees—this is going to become one of the most contentious privacy issues in the 21st century because it constitutes such an egregious violation of not just privacy laws but search and seizure law.

Beyond the discovery of drug or alcohol abuse, there is the potential risk that testing could reveal the existence or affinity for specific genetic conditions—and that will open a host of privacy issues that could level a company to the ground. What genetic markers is the lab allowed to screen for? Under what provisions of the law? What will be made of the findings? The only occasion in which genetic testing has been proven valid is environmental toxicity testing. But there is more to this than the law. If an employer discovers a great employee has cancer, what do they do with the findings? If the person is still performing their job very well than it isn't relevant to the position or their performance. It also does not take into consideration a fact of life, just because they are terminally ill does not mean the rent is free—they still have to work. That is unfortunate but it's the way our society functions. When it comes to any kind of testing, no matter the reasoning, employees should immediately consult their respective government office and their attorneys; and employers should do the same long before they ever take that step.

Much of what I've covered so far would fall under the rubric of intentional surveillance but what happens when management discovers things through the grapevine—a form of unintentional or at least passive surveillance; a case in which management is just keeping an ear to the ground and it would be safer to assume this is the norm and not the exception. So then what happens when a manager gets wind of something going on but that is of a clearly personal and private nature—it seems to depend on what it is and what impact it has on the job.

It has to start with the understanding, above all else, that sometimes bad things happen and tragedies befall all of us at some point and there most certainly seems to be an increase in stress related problems such as alcoholism, marital problems, clinical depression and everything in between—that this occurs isn't really the issue.

Without getting carried away with cases it helps to understand how delicate these problems can be. If you're a manager, how would you react to hearing that one of your "Super Producers" was seen drinking more than usual lately at a bar near work? If there was a direct relationship to their performance there may be reasonable grounds to talk to that person but barring that, how could you confront something that was not having an impact on their work? Are they drinking more because they are stressed or hate their jobs, or is it more complex than that? There is a host of other problems but here are two more.

Marital problems rank right up there in terms of stresses that affect a person at multiple levels including their performance at work, and the common myth of 50% isn't far from the actual divorce rate in the United States—the good news is divorce rates have been declining since 1981 and hover in the 40th percentile according to the Census Bureau, the bad news is that this doesn't not take into consideration separations or other forms of marital dysfunction. In whatever form it takes, no matter how you spin the "family values game", it is a sobering reminder that if you are between 25-35 there is a 33% probability that your first marriage will end in divorce. Now compare that to statistics for 1880 in where divorce rates hovered at 5%. So, assuming you take this at face value, it is obviously going to have some impact on work performance. No matter how much may be said of productivity, it's pretty hard to focus on yet another report or a meaningless meeting when someone is losing their house let alone their partner. The privacy angle stems from how management treats these individual situations—different companies will have different strategies with some being more compassionate than others.

Alas I have saved the best for last—the notorious office relationship that carries with it more baggage than a 747 not to mention the kind privacy, personnel, organizational implications that give managers nightmares, and hereafter reveal my "true" colors on privacy. To place this issue, and take note it is not defined necessarily as a problem, in the right context requires a least a basic appreciation of its prevalence—a fact confirmed through statistical data rather than a hunch;

that according to Kinsey, 60% of men and 40% of women have an affair during the course of their marriage. And to tackle the issue head on further requires that you first accept its commonality, secondly to accept the prevalence within any given company, and finally to put aside your "moral" convictions and focus on the privacy angle.

Whether driven by lust or love, whether married or single, relationships do develop in the workplace and to rely on simple misconceptions does not do justice to its depth and complexity—and if you are on the management team it is crucial that you fully understand the push and pull forces that cause it to happen in the first place—long before instituting awkward, draconian, policies that you cannot enforce anyway. This concept of push and pull forces indicate a combination of internal and external forces that guide a person to act upon these wishes. Anyone can be *pulled* towards an office relationship for a variety of factors including a natural attraction, curiosity, risk, and genuinely falling in love—but internal forces may be providing the precursor conditions or the so-called *push* that can be caused by virtually any number of conditions such as boredom, to fill gaps within an existing relationship, and finally what I suspect are the primary factors, either the need to escape or the need for attention; it is highly possible, even likely, that it is *more than one* of these that push a person from their relationship and pull them into another.

Whatever it is that causes these situations to develop, how a company deals with it is acutely telling of its attitudes not only towards its general position on employee privacy but the organizational attitudes towards human dignity and maybe even the management team's intimate understanding of human nature—are they out of touch or do they truly "get it"? Of all the organizations I've studied the most compassionate appear to take a discretionary don't-ask-don't-tell position so long as the relationship does not interfere with the company—it is to put it mildly a relatively benevolent approach and in my way of thinking the only way to handle it. Still a caution is called for those Human Resources Managers and organizations that might otherwise want to "stomp" on this "problem", you can't mess with Mother Nature and it is a horrible intrusion on the most intimate dimensions in a persons life. Biology will always prevail over childish attempts to curtail it.

7.4 Best Practices for Consumers

"Peace is such a precious jewel that I would give anything for it but truth."
—**Matthew Henry, *English Theologian* (1662–1714)**

While there is clearly a need for increased constitutional and statutory protection of privacy coupled with improved business practices, the fact remains that consumers have an obligation to protect their privacy as best as possible. It will not eliminate privacy violations but you can do a lot to mitigate the damage before it ever happens to you or your family. To that end the best practices compiled and discussed are here to do with as you see fit. If there is a single word to describe a new strategy for you it would have to be "vigilance"; by being much more vigilant and demanding you stand a better chance of preventing others from abusing that which is rightfully yours to begin with.

As a prime example of this, there was one occasion that I recall vividly where I wanted to purchase some equipment and intended to pay cash, a presumably straight forward transaction, but the "system" at this particular retail required a customers phone number and mailing address to close a sale. This wasn't the clerk's fault—the system, and by extension the companies' policy, was to collect personal information for marketing purposes—and designed the system in a manner that coerced customers to provide—needless to say I politely cancelled the purchase, walked two doors down and bought the same equipment without a problem.

The maxim is, whenever possible, provide the minimum amount of information required to complete a transaction—and be willing to refuse it; some have argued that it is better to use case at all times but this is unrealistic and inconvenient and even if I was to suggest it I am not sure that many can resist their cards. You might also consider asking the company for their privacy policy and ask whether any personal information is going to be shared, exchanged or sold to any third party or affiliate. All relationships, whether personal or commercial, are predicated on trust—if they are going to exchange your information, it is a violation of that trust.

While I am on the topic of information requests, your credit rating and the information held by credit bureaus warrants special attention, after all every bank and lender are making crucial decisions that affect a person's life based in large part on the rating that has been assigned by the creditors that you've historically been associated with. Based on the concepts of Risk Based Pricing, lenders could be basing their decisions entirely on a history that could be incomplete, out of date, or worse, inaccurate. Even if you were granted credit, anomalous entries may result in higher interest rates than you should be paying. Beyond lending decisions, a credit rating is being increasingly used in decisions that affect education, employment, and housing.

From privacy perspective the use of credit ratings in non-credit transactions is becoming a very worrisome trend, especially in the insurance industry where there is this school of thought that the level of a person's credit rating is in some way proportional to the degree of risk they present to a carrier. Though there exists no casual link to substantiate this correlation, it is being used nonetheless. There has been harsh criticism against the practices that have been deemed an unfair business practice and some states are beginning to tackle the problem by banning insurance companies from basing decisions on credit; ultimately they shouldn't have access to it at all. It's about time.

Inaccuracies in your file can also cause considerably more harm than you may realize and it is imperative that the credit file be checked at least once a year to resolve problems, outdated entries and other inaccuracies that could be affecting your ability to acquire positive lending conditions. And if the study from National Credit Reporting Agency and the Consumer Federal of America is any indication, over 29% of all surveyed individuals had significant errors in their files. And, as unfortunate as it is, there are sufficient studies that show women and minorities are being discriminated against based on their geography and other personal information that has been provided such as age, sex, marital status, and race. I would love to say this isn't the case at this point of history but the study by the District-based National Community Reinvestment Coalition confirms this discriminatory practice. Different scoring practices are used by the predominant credit bureaus but all of them use a form of ranking that determines your credit worthiness; a FICO score ranges from 300 to 900 with any score over 620 considered a prime rating. Canada uses a different scoring algorithm that uses letters instead of numbers but the effect is generally the same. The sooner you check your number the sooner problems can be addressed.

When dealing with people over the phone or through mail, be even more cautious than ever. A simple measure to cut down on the annoyances and the invasion of privacy is to get call blocking and demand that you be removed from telemarketer's lists—don't ask, demand it. If the operator refuses, speak to managers that have the authority to do so. You can also get yourself removed from many lists by calling the Canadian Direct Marketing Association in Canada and the appropriate Do-Not-Call Registry in the United States. You would also do well to avoid 800 and 900 series numbers like the plague and always hang up when computers call. Automated call centers are simply calling millions of numbers using a computer which reduces the labor required to find "live" targets—if you wait on the line you will be flagged as a target.

Of all the things that ought to turn on your radar it is those special offers, "Free Kits", contests, and surveys—every one of them has the single goal of collecting information about you and you are implicitly consenting to unknown uses just by providing it to them. The same can be said of convenience cards and any form of "loyalty" schemes—just keep in mind that whatever you give is going to be shared or sold a hundred times over.

Before I'll finish this section, I would like to also talk a bit about Transactional Information Relevance. This fancy term refers to what information is absolutely required to complete a transaction. You must establish what you are willing to give up in exchange for something you value and determine whether the request is valid. No matter what you are buying, with few exceptions like insurance or medical services, there is absolutely no need to provide any information about your spouse, children, beliefs, education, or anything else—if you feel they are crossed the line, then they did. If I bought a pool, the supplier does not need to know who's going to be using it. Do yourself, and you family, a favor by writing down what you are willing to exchange and under what circumstances; in other words, develop your own privacy protection policy—and then stick to it. Remember that you are not only limiting your exposure to businesses but you are also limiting yourself to identity theft. Shred mail before throwing it out, the same for bills, invoices, credit card statements and anything else that would be of value. And if it is valuable, buy a safe or lockable file cabinets.

When in doubt, especially when confronted for demands for personal information that appear inappropriate, it may be necessary to determine if the privacy laws of your state or province cover such questions. In Canada, we covered by several layers of law as it was noted in Chapter Two, but U.S. laws tend to be very chaotic and what is covered in one state is not always covered in others. Derived from the Compilation of State and Federal Privacy Laws (1992 ed.), most states protect against unwarranted disclosure of arrest records, but Arkansas and Alabama do not provide this protection. To look at Alabama's record on privacy protection, it is obvious that privacy has not received the attention or ally it rightly deserves; there is no protection for credit records, employment or insurance records, there are no privacy statutes, nor is there any protection for social security numbers. That is not to single out or show preference to a given state, every state had varying coverage for different types of personal information records. You will have to inquire for your own state to determine if you are concerned that personal information is being collected illegally.

For all that has been said, for all the advice (much of it common sense), it all comes down to using your judgment and learning to say "No". But to gather the strength to say "No", you must first come to terms with the simple realization that everything except your name is private—and anything else you provide ought to be viewed by an organization is a privilege not a right. It doesn't matter what they think they need or deserve, you are the sole judge of what you are willing to part with. In short, the objective is not to make you paranoid but just to get you to think about what you are giving and to whom.

7.5 Best Practices for Business

"In the business world, the rearview mirror is always clearer than the windshield."

—Warren Buffett (1930–)

Far be it from me to dictate to businesses how they conduct their affairs or set forth for them the philosophical foundations that define their ethical standards. However, for those that strive to improve upon their privacy practices, what lies ahead are generally accepted privacy practices that fall under the rubric of fair information practices. The best I can provide the guidelines for developing a meaningful, coherent privacy policy that will engender trust with your customers and improve customer retention. Mind you, from the highest levels of management, must come the recognition that customers are finite and invaluable to the survival of your organization and the corollary recognition that private information belongs to the customer. It is their right to decide what will and will not be shared with a company no matter your rationale. If you truly believe in the expression that the "Customer is always right", then this should not be that difficult. What follows is a series of questions that will form the basis of policy development and review.

What measures have or will be taken by your company to safeguard personal information?

Beyond the issues associated with corporate security, the protection of customer information has privacy implications that are becoming increasingly driven by legislation and the potential threat of litigation; and the steps an organization takes to protect information about its customers and employees are something that needs to be addressed at both the strategic and operational levels. At a strategic level, every organization regardless of size or line of business must develop and *continuously* review an effective Security policy that covers physical security, organizational security, and information security. At an operational level, enforcement of this policy is just a matter of implementation; it is a matter of embracing privacy as a core principle in every customer interaction.

Physical security refers to the measures a company takes to secure facilities against theft and intrusion, particularly in sensitive areas. Every area of the company must be examined using objective criteria to determine what must be done to protect against all types of threats. For example, what has been done to prevent access to the key entry points in the building such as the front door, fire exits, windows, roofs, and ventilation systems just to name a few of major points of access. Do the company use access cards in different areas of the building, and what are the criteria used to grant access? This is a small subset of the questions that need to be answered by management.

Operational security is all about processes and people. It is up to every company to determine how information is shared with employees or suppliers and it is equally must be determined who has a "need to know" as well as vetting the staff to ensure they are trustworthy which is largely dependent on the level of access they require. The degree of severity depends a lot upon the nature of your business and management must consider the risks of exposure, espionage, and theft.

Finally, and the focus of this question, is information security; after all, there are many preventative measures that can be deployed and may well prevent some of the problems previously noted. At the very least your I.T. Department should be involved in the process to determine how information is maintained, stored, and protected during and after it is used. Who has access to customer (or employee) personal information? Do you effectively use auditing to determine a history of access? Does the company deploy encryption with sufficiently strong key management practices?

It is strongly recommended that firms consider recruit external consultants or advisors to conduct these audits as well as an internal audit; by using independent advisors, management avoids the risk of coming to the conclusions they want to hear as opposed to the problems that may exist. In smaller organizations, there are privacy advisors that can provide cost-effective audits while larger organizations will require a more extensive audit performed by the larger consulting entities; that is not to imply that smaller firms cannot conduct an audit, but admittedly larger firms have multidisciplinary teams that include attorneys, technology advisors, and a host of other experts and they are quite familiar with regulatory compliance.

Is your customer and employee information accurate and up to date?

You do not need to be Warren Buffett to realize that the profitability of a corporation is based, in part, on the accuracy of information that is used to make strategic and operational decisions and the cost of errors in company databases can cost an enormous amount of time and money to correct; if this is the case, then it begs the question why companies regularly allow inconsistencies and errors to persist in organization data sources. Privacy audits aside, conducting periodic audits and reviews can mitigate a majority of the problem.

Does your company limit the collection of personal information?

Beyond the statutory limitations and penalties, is your company collecting more information than it really needs to effectively conduct its business? It is not unusual for companies to ask, retain, and store tons of data that has absolutely no impact or relevance to the line of business; the more indiscriminate your collection efforts, the more likely that you will alienate your customers. That is not to say you shouldn't try to understand your clients, just limit what you are asking. If that isn't enough, consider that the less information you collect the less the cost of managing it and reduces the risk that it will be abused.

How is this information disclosed, retained, and used?

It is awfully tough to ask an organization to purge data that they feel may be needed at some point in the future but there is value in reducing the amount of data that is retained and made available; even if the data cannot be deleted due to regulatory and accounting compliance, that does not mean the data can't be off-loaded to archives; for example, an annual audit might move the years previous data to a different server or to tape. The greatest concern to consumers, though, is the unauthorized trading or selling of customer-centric data to third parties not known to the individual or involved in the original transaction. More than any-thing else, adopting deceptive marketing practices that obscure the fact a com-pany is sharing information is a sure fire way to alienate consumers. If the social and ethical concerns are not enough, then consider the economic imperatives. Providing information to third parties says a lot about a company and very few can afford to anger the public.

To what extent would you rate your company in terms of accountability?

If an attorney was to file suit for violation of privacy or your firm became the object of an investigation, would your organization be able to show that privacy was a senior management priority and would there be sufficient *evidence* to show there were active processes in place. Could you or senior management demon-strate adherence to fair information practices? It is recommended that organiza-tions assign, retain, or contract a Privacy Advisor with sufficient management authority to intervene in the fact of conflicts and decisions made at every level. The company should also develop a comprehensive privacy policy and the pro-cesses to assure enforcement.

Does your company openly ask for consent before information is collected, used, or disclosed?

Above and beyond the legal requirement in many countries that companies must seek consent, it is imperative that the reasons for collecting personal information be clearly communicated to customers and employees before the collection actually occurs. Under no circumstances can any company compel a customer to provide information by implying the withholding of services or goods or employ deceptive marketing practices in its collection. As noted in Chapter Two, at least in the case of Canada, consent for the collection, usage, and disclosure of personal information is now mandatory.

Has management clearly determined the scope and purpose of the information that is collected?

It is imperative that every organization identifies the reasons for collecting personal information and the scope of that information for the purpose of conducting business. In a cash-based business there is no need to know anything more than whether the customer has the ability to pay. For example, despite efforts to the contrary, gas stations do not need to know where you live or work to provide gas. There is a concerted effort to retain customers by offering loyalty cards but customers are not compelled to provide information in exchange for many services or goods. This will depend on the transaction—credit transactions need information about a consumer's employment history and credit rating to make decisions about who is entitled to credit. But the company does not require information about the customer's children or where they went to school.

Does the firm simply the process for the resolution of issues regarding information held by the company?

How does your company contend with complaints and concerns raised by customers and employees? By having an open, responsive communications strategy, an organization can demonstrate a degree of transparency that can be translated into improved customer retention and an improved public image. The company ought to have a concise, unambiguous policy to deal with how your customers interact with you, how they can get access to the information concerning them, and what remedial measures are available to rectify errors or discrepancies. Keep the policy open, clear, and simple. To make things all the more effective, complaints should be immediately conveyed to senior management in an effort to prevent problems from escalating—dealing with it now is much better than dealing with it in court.

By asking these key questions, the management team, in cooperation with appropriate counsel, can circumvent many problems long before they get out of control—and it will serve as a means of addressing the primary focus points of privacy legislation. The more proactive the policy the less change the firm will have to endure.

7.6 Practicing What We Preach

"Do not let your deeds belie your words, lest when you speak in church some-one may say to himself, 'Why do you not practice what you preach?'"

—Saint Jerome (374 AD–419 AD), Letter

Much of that which follows in this chapter tends to focus more upon a variety of privacy issues, but that are connected by the common threat of practicality; though you may never have to deal with biometric scanners or have to worry about genetic testing, issues such as identity theft and workplace privacy are a lit-tle closer to home. In addition to these two key topics, the rest of the chapter moves into pragmatic recommendations and best practices for both consumers and the private sector in an effort to improve your privacy practices regardless of whether you are an individual or organization. Just reading it alone will change absolutely nothing whatsoever—it requires actions on everyone's part that will prove inconvenient at the very least; and like any regimen, it will take some time to get accustomed to. There is nothing here that forces you to adopt any of the measures that are provided, that is purely a matter of personal choice and corpo-rate culture but they are provided in the hope that it will, at the very least, pro-vide a foundation for improved privacy.

I had originally debated the merits of discussing identity theft but was over-whelmingly swayed by its relevance to privacy and the catastrophic impact that it has upon its victims. There are far too many people that continue to be robbed of their name and livelihood to leave the matter untouched and that alone justified its inclusion; it is much better to give you fair warning now than to discover you were victimized tomorrow. That identity theft has emerged as a high priority with law enforcement, coupled with the increasing number of incidents that have been reported, should be ample warning that this crime has no boundaries nor does it discriminate—to "them", it only matters that your name can be used to defraud you or at the very least hide themselves. That your life is left asunder is of little consequence. The challenge for me is to, somehow, persuade you that you are vulnerable and that countermeasures are necessary to reduce the risk that you'll be targeted. Ultimately, the less information that is made available to the public, the less information that is open to abuse and the countermeasures that have been provided for you will at the very least reduce of privacy invasions or identity theft. I won't pretend, nor have I ever, that the risk will ever be eliminated in full–but better to be reasonably secure than not secure at all. In the current state of affairs, it would not take more than a days worth of intelligence gathering to compile a reasonable dossier on any target of my choosing. The less information a thief has the more likely he would move on to an easier target.

The debate over workplace privacy is much more volatile than of identity theft; in the case of identity theft, no one would argue that there is a rationale for theft of one's identity. In the case of workplace privacy, however, there are diametrically opposing views on the subject and you may assume correctly that there are those that oppose all forms of workplace invasions and, conversely, the employers, not many but some, that counter that employees are no different than any other form of property. I do not believe, not even for a moment, that a majority of employers take the view that you are of little more value than a desk; on the contrary, I would expect that most employers understand the limits to which they may invade the privacy of their staff, and somewhere in the middle of this skirmish line is you. My position on this matter, as it has been sensed throughout the other chapters and stressed repeatedly, is that there exists no room for extremist positions on either side of the argument. There must be a "middle ground" in which both employees and employers will find an acceptable compromise between expectations of privacy and productivity. First, it must be understood that the assets owned by any organization were not acquired for our collective amusement or personal use; every organization exists with a mission to achieve, and distractions from the objectives set forth do little but diminish the probability of economic survival let alone success, and by extension that includes our success. Suffice to say, companies do not invest millions in their phone systems so that we can socialize to no end nor do they invest millions in Internet and e-mail access so that employees can share pictures of their kids, to say nothing of the "other" type of pictures. Despite my positions on privacy, they would find no ally in me. That does not begin to address the arguments with respect to labor productivity—employees are duly compensated to perform a specific task or tasks that further the goals of an organization. Therefore "excessive" attentions to personal affairs sabotage that mission.

There is always a "but" and little doubt you saw it coming. There must exist a reasonable *boundary* beyond which an organization, whether private or public, simply dare not explore. To track every movement, to listen to every phone call, or log every keystroke is symptomatic of an organization that has lost sight of human dignity; just because you *can* do this to your staff does not infer you *should* be doing this to your staff. It is further symptomatic of management insecurities in the same manner that micromanagement is equally symptomatic of this problem. At the very least you, as an owner or manager, have a moral obligation to clearly notify staff when and where they are monitored. There is, sadly, *far* too much evidence to deny the existence of the problem; too many lawsuits, too many investigations, and too many media accounts for any employer to categorically deny it has ever happened. There was once a need for labor laws to counter abuses, and now there is no less a need for similar protections against violations of employee privacy. Biometric devices and smart cards are being used to track and predict movements, one-way mirrors let supervisors watch every movement of their charges, and closed circuit televisions speak for themselves.

For all of the guidelines and recommendations that have been noted, the fact remains that staying on your toes is the best practice of all; the more vigilant you are, the less likely you will find yourself surrendering your privacy. If that should come to pass, it is much more difficult to undo the damage, and in some cases it's impossible. The law is finally beginning to show its fangs but it does not absolve you of the duty to take responsibility for yourself—and that includes the fortitude to say "No". Before you can say "No" to a merchant you'll have to first learn to say "No" to yourself. I am not, in anyway, decreeing what constitutes acceptable behavior, I am too much of a rogue to interfere in your affairs, but I am making suggestions that are left to you to act on. Do with them as you see fit but at the very least learn to think twice.

Private enterprises, in light of new legislation in many countries, have much less of a choice as to their privacy practices and it is incumbent upon every organization to understand the new laws, especially in Canada and the European Union, and implement these practices as required. You have little choice but to begin the arduous process of auditing your information collection, retention, storage, security, consent, and disposal practices—if you don't audit them, you may well find yourself the target of government audit. Rather than lament this overarching legislative sledgehammer, it would be more productive and perhaps profitable to leverage on fair information practices that will improve customer loyalty. After all, selling information to a supplier or third party was never part of the transaction and is now, as a matter of law, a much more difficult proposition given the need to gain consent from *every* customer. Show a customer that you take their privacy seriously and you'll gain more loyalty than any advertising campaign.

What we say and what we do is inversely proportional to the difficulty in applying the principles that have been espoused; the harder it is to make changes in behavior, the more likely it will remain in the domain of lofty writers and the meaningless banter of academics. I never said it would be easy to change certain patterns of behavior that would deprive you of certain conveniences. Letting go of that convenience card or not participating in that great giveaway for that new car can be equally not easy. But failing to think critically has its costs too—costs that are not always visible or apparent. My wish is not that you change every single habit you have nor deprive you of what the market has to offer, I would only hope that you are able to appreciate and give more thought to what it is that your surrendering. In the case of identity theft, the actions of the individuals and the vigilance maintained will determine the probability of becoming the target of an identity thief. Even in the case of workplace privacy, it is the responsibility of both employees and employers to establish what is and what is not an acceptable code of conduct and surveillance. If there is indeed such a loud chorus extolling the virtue and importance of privacy, then it is about time everyone began to practice what they preach.

8

Conclusions

"Where there world ceases to be the scene of our personal hopes and whishes, where we face it as free beings admiring, asking and observing, where we enter the realm of Art and Science."

—Albert Einstein (1879–1955)

Well, for all of that reading, for all of those philosophies and paradoxes, after all this time, one cat is still chasing his tail, blissfully ignorant of the fact that it is his, and the other one continually attempts to crawl into my lap despite the obvious fact I am trying to get my work done. Though they have been unwittingly employed as an opening and closing metaphor to privacy I doubt they are going to change anytime soon—you, on the other hand, possess the ability to do just that. Chapter Eight is conspicuously shorter than the others. Do not, however, let its brevity lead you astray; it is not only the last chapter but also the last opportunity to persuade you to take your privacy more seriously. You don't have to agree with the things I've said, or for that matter anything I've said, but if it does make you think a little more, than I must content myself with that.

There is at least some degree of confidence that approximately 75% of readers would agree they are concerned about their privacy; an assumption that is supported by the numerous studies that were presented earlier. Nevertheless, those are responses from the general public that have likely never read a book on privacy; it would be interesting to re-poll them after reading this book to determine whether that undecided 25% change their mind at all. It absolutely makes me wonder who that 25% really are, what do they do for a living, are they staff or management, what demographic groups do they fit into. If you were to ask ten friends the same questions and two said they didn't care, chances are you would be more curious about their thinking than the other eight.

I am still inclined to model the privacy issue as warring factions and for that reason there is the temptation to classify people into several groups. The first group may be referred as the so-called Watchers and Controllers that have a vested interest in invading a person's privacy for whatever reason. It could be for money, power, whatever the case, it doesn't matter—they seek intelligence about others with the intent of modifying behavior. The second group is the vast majority of the public that either lives in blissful ignorance or remains undecided on the value of privacy or freedom. And lastly, there are the Guardians, which I divide into those that are actively engaged in protecting privacy and those that have sufficient knowledge not to fit into the second group.

A majority of you that read this book will fit well into the third group for the simple reason that you now know far more than most and will develop a stronger sense of situational awareness. Naturally, it is up to you to figure out which group you prefer and realize these classifications are arbitrary—I could have all sorts of scales or models with differing degrees of granularity. Then again, if this was a "perfect world" then it is safe to assume that 100% of readers and respondents would have fit into the third group but every reader knows this is not the case and, at best, utopia is, perhaps, a state of mind if it exists at all.

Much like privacy and freedom, there are so many conflicting priorities, interests, external influences in a person's life that tend to result in highly contradictory behaviors that stem from the difference between how we would like to live and all of these factors that prevent us from living according to an arbitrarily idealistic view of the world. It is relevant to privacy and freedom and you can even find relevance to virtually any aspect of life that you choose: religion, leisure, politics, marriage, all of these things seem to exist in a constant state of tension between idealism and realism.

I cannot stress enough how important this idea of a paradox is and it is one of the reasons it was introduced so early in the book; granted I might have gotten a little carried away and introduced a few of the better known paradoxes that you either solved or cursed me for—they had a slight relevance to privacy but they can be a lot of fun and they do emphasize the point that things are not always as they seem and there are infinite parallels in human nature to explore. We, collectively speaking, have a strange habit of saying or believing one thing and doing another.

For example, at the deepest of personal interactions, according to Social Penetration Theory, there exists a contradiction between the desire to disclose intimate details with another person, presumably one that you trust over time, and the paradoxical desire to withhold information. Based on this theory, disclosure is seen not only as healthy but it is actually necessary. As it was noted in Chapter Two, the degree of trust, and my own theory of trust equilibrium, play a significant role in determining the extent of disclosure. That trust, though I have spoken of it so much, is really the gateway to facts that would otherwise be unspoken to anyone else, or in other words, trust is the gateway past the internal guardians of a person's privacy.

It is also a paradox because on the one hand I might want to know more about you and presumably you of me, but I don't know you well enough to do so yet. Somewhere, in the course of a relationship, one or the other or both must make a decision—that the person sitting across from you has your best interests at heart and would not cause you harm. Even as I have discovered, once in a while there is that rare person with whom I'd find it difficult not to share my deepest thoughts with, and they know who they are, but the point being they are rare, in fact I can personally only think of one or at best two.

It is too late to be getting into other paradoxes that come to mind, but before getting back to privacy, work tends to present fascinating contradictions and if you are a manager, owner, or executive of a company, this is going to be hard if not impossible to accept. There is, nonetheless, a paradox that exists between how people really think of their jobs, and work in general, versus the perceptions that we put forth. Statistically over 75% of those surveyed hate their jobs, and it crosses most demographic boundaries—in other words the data seems to indicate a universal dislike. And it is this contradiction of internal and external forces that makes morale boosting strategies quite challenging. Obviously there are external forces, notably economics, that compel a person to work—but that same majority would much prefer to be golfing or going out with their friends—and as jobs continue to encroach on the home and time spent with families I am of the conviction that this paradox will continue to polarize positions all the more and that explains, partially, the push for balance.

That aside, attitudes and behaviors towards privacy and freedom exhibit similarly paradoxical behaviors; studies certainly show a majority of the population that is deeply concerned about their privacy. If that is the case then how does one account for this willingness to recklessly sacrifice something of great importance for things of questionable value—it doesn't compute. The same paradoxical behavior shows itself in the context of freedom and civil liberties. Despite a fervent belief in democracy, how many people have actually read the Constitution let alone the Patriot Act? I would suggest the number is remarkably low. Have readers come to the understanding that privacy is directly related to freedom and that you cannot have one without the other?

In an effort to explain this conspicuously contradictory phenomenon, there is a chasm between the idealistic vision of how the world is *supposed* to work and how it *actually* works. In a perfect world there wouldn't be a need for this book in the first place because there would be no concept of invasion to justify its existence, but then again, the world would also be free of poverty, war, and famine although five minutes of CNN will alleviate that misconception very quickly. If this was indeed a perfect world, there would be no need for draconian privacy laws either; but attempts to allow for self-regulation and self-control have failed miserably and this shouldn't come as a surprise—when given the opportunity to serve our self-interests versus that of the greater good, which do you think will prevail. You may be able to put the interests of others above your own, but that is not a universally held belief and until it is, we need protections to guard against abuses.

At the risk of regurgitating my thoughts on all things technical, the same protections against abuse must be extended to science and technology to provide at least a semblance of assurances that new developments will be constrained by limitations on their use—it doesn't matter how great the new gadget is, it doesn't matter what potential applications it may be capable of, the boundaries of privacy and freedom must be taken as inviolate. That is going to take a lot of political will and the only way, to my mind, of doing so is through media scrutiny and public protest—once an abuse gets press and the media chew on it, there is no way to push it back in the shadows. The looming threat of outsourcing jobs would not have received the same degree of attention at a political level if CNN had not given it so much air time. That public exposure was partially responsible for pushing that issue into the presidential campaign.

Still, I am concerned that the negative aspects of progress and technological development have clouded the judgment of the public and little attention is devoted to the positive benefits that technology can provide. Just as I am not inclined to believe in the mating myth of Mister or Miss Right, technology is neither perfect nor evil but rather has the potential, like people, to be either one or a little bit of both. People, myself included, have many good qualities but there are also those that I am not proud of—and dependent upon the person, they will either be predominantly good or bad. Without getting into it again, genetic testing is by far the most controversial of sciences and possesses the ability to save millions of lives and yet it holds the power to destroy the most sacrosanct right to personal and bodily privacy and integrity.

Putting aside these philosophical and theoretical musings, I'd like to believe that the suggestions and recommendations will be taken to heart and implemented when and how you see fit to do so. At first glance, many of these recommendations and guidelines seem to have absolutely no bearing on your privacy but they are; everything from the proper use of e-mail to the installation of a firewall make it increasingly difficult to acquire personal information from your systems. Understandably, applying all of the principles mentioned in Chapter Three will improve the security of your home and prevent the onslaught of viruses, Trojans, and spyware—they are worth putting into play either way.

All of the firewalls in the world, though, will not do a smallest bit of good if it is not backed up with a healthy sense of skepticism, self-control, and common sense; these three elements are worth more than all of the firewalls in the world and technology cannot replace them or make up for their absence. Despite this warning, this constant state of vigilance can be difficult to maintain and requires substantive changes in a person's behavior that will take time before they become second nature. This need for vigilance becomes all the more crucial in the presence of your children that do not yet possess the critical thinking skills of an adult.

Oddly enough people would never, at least where I live, think of leaving their house or cars unlocked but take little in the way of precautions when they are online or at the mall; and that makes me think that all of us could use a primer on operational security assessments that we use in the computing field. These assessments tend to look at all aspects of security and not just the computers or networks; it wouldn't do any harm to rethink your physical security as well as your operational security. In the former case I am referring to everything from your house to the papers you keep—do you have fireproofed lock boxes or a safe? Are your most important documents, like a passport, secure from theft? Do you have a safety deposit box to keep critical paperwork that should never be kept at home? There are so many things to think about. Operational security focuses on what you do rather than what you have. Turning off unused services, frequent credit checks, verification of change of address—all of these things will help in preventing identity theft or at least minimizing the likelihood of it happening—yet from my own observations, I do not see sufficient evidence to indicate that these practices are used as much as they should be. Never said it would be easy, never implied it wouldn't involve changing behaviors and habits—but they are still better than doing nothing.

Putting aside the more pragmatic components and recommendations contained therein, you are of course entitled to ask to what extent this book is a work of fact or fiction and more to the point, whether the author is a foaming-at-the-mouth leftist social-activist that has more causes than jobs. I'd prefer to think of myself as a very moderate liberal, but frankly my personal convictions are not important in the grand scheme of things.

What is truly at issue is the question of whether the technologies and dirty little secrets I have "exposed" are true or not. Well, without question, all of the technologies that I have brought to your attention, none of these are a figment of the imagination and there is a wellspring of publicly availably information both online and in the libraries to corroborate everything that I have set forth, provided of course that anyone has the energy to pour through the documentation—you are most welcome to do so. I am not the creator of this "stuff", but only a messenger.

Even if you aren't sure, the evidence for the existence of the Echelon surveillance network is not the product of an over-active imagination and several of the participating countries have already, much to the chagrin of the United States, come forth and have acknowledged its existence. It is also hard to refute the existence of the Total Information Awareness systems or Carnivore when the Department of Defense the FBI both testified in Congress. Of course if you aren't sure, you can always visit the FBI and read its own statements. GPS, biometrics, genetic testing—none of these technologies are the figment of imagination.

However, it is the application of technology and whom that technology benefits that ought to be called into question. Obviously those that invest in new science and technologies should be able to seek a return on their investments to offset the inherent risks of research, with the proviso that such developments do not cause harm or deprive people of their liberties. I have no issues with the development of a new car—I have issues with a new car that can be tracked covertly without your consent—there is quite a difference between the two. Likewise, I have no problems with next generation cell phones unless evidence suggests the new phone includes the capability of locating me without my consent. The challenge is not only the technology but also this unusual willingness of the public to "buy into" whatever the public relations people pull out of their hat next. Your call is being monitored for quality purposes.

There have been a number of occurrences where I have been willing to put forth what amounts to wanton speculation and musings on the future; even if they are reasonable speculations they are nonetheless educated guesses. Nevertheless, wherever possible the wording has been phrased in a manner that differentiates between what is clearly evidentiary and what is my opinion—and in so doing it far more honest than stating an opinion as fact. There is nothing to say that these musings of the future are accurate but then again there is no evidence to suggest that they aren't. That is a matter of subjective interpretation and you are welcome to accept or challenge my logic. To believe or not believe is a choice that is left up to you and it is not my place to interfere with that process.

I have my doubts whether it is ever easy for an author to finally stop the writing and editing but that time has come; so it is only fitting that I should bring the discussion full circle to my seemingly obsessive quest for balance and equilibrium that I have managed to interlace into virtually any topic that comes to mind. What is it about this principle of balance that is so inextricably tied to privacy and freedom?

Though recalcitrant and unapologetic in my opinions, it has crossed my mind that what I am really trying to convey is the need to establish a line of control between what is private and what is public; or if you prefer, establishing a clearly defined boundary over which no organization, government, or individual may cross without one's express consent. Regardless of the nature of any given interaction or transaction, there are just some things that must forever remain off limits with few if any exceptions.

Needless to say, there will always be a need to communicate personal information for the proper functioning of both the economy and society and at no point have I advocated abolishing the collection of all personal information; if anything, that position alone, places me into the more moderate schools of thought. How can any commercial entity be expected to grant credit without a reasonable understanding of who you are and your ability to service your debt? And even in the presence of data, what is really relevant anymore—there was a time when people would spend a better part of their lives with one company and now the average is three—four years and approximately six career changes in a lifetime. Having a base of information such as where we work or how much we make coupled with a history of credit worthiness is perfectly reasonable.

The boundary I keep alluding gets crossed when companies or governments begin to collect and combine information that intrudes the privacy of our homes, family, and persona. It is one thing to deny a loan to a person because they do not make enough money and something altogether immoral to deny the same person because they drink or smoke on the grounds they might die or that this perception of risky lifestyles in any way speaks to a person's financial health. As well, it would be no less immoral to base financial quality of life decisions based on the results of drug testing or genetic testing—that is only going to bring us down a very dark path.

There is a sense that there are improvements underway, at least in some countries like Canada and the European Union, where tough privacy legislation has come into effect and is only beginning to be felt. So far it would appear many if not most companies have not even begun to contemplate, let alone comply with the new laws and it will take a very long time before this transpires; but it is a positive start and it does provide a degree of confidence that these countries are beginning to truly take privacy seriously. The private sector will ultimately have to re-engineer its business processes policies over the next few years but we have to start somewhere. I am still of the opinion that many companies really have no interest in the private lives of their customers or employees and it is the few intransigents that have made for bad press. Nonetheless, the rule of law dictates equality before the law and that means universal application of the law—everyone has to comply regardless of his or her practices or policies.

My concern in writing about privacy and speaking to the human or social dimensions of privacy is that you may get the mistaken impression that you shouldn't maintain intimate contact with your friends—nothing could be further from the truth. The very survival of the species, for that matter your survival, in part rests upon the ability to communicate and share with others—we are, after all, social animals. Curiosity, gossip, and secrecy—all of these things are natural and healthy components of the human condition and I wouldn't want you to think that you should suddenly crawl under a rock. If anything I would say the opposite is true—that amongst a core of your closest and most trusted friends that a little more disclosure might even be good—that, though, is your call. At the very least, having a better understanding of these things will help in understanding why people do and say the things they do. Whether in the course of commerce or personal relations, I have to return to the right to be left alone; that this represents the bedrock of freedom and the right to live according to our own choices and preferences. Ultimately, a person must be able to live free of interference in his or her affairs and that personal information be regarded as property of the person to be used in our own self-interest. There are already laws for protecting intellectual property so maybe it is the right time to start thinking about protections for our DNA.

978-0-595-38843-1
0-595-38843-4